H. R. Kirby
2, Selwyn Drive,
Eastbourne,
Sussex.

THEORY OF TRAFFIC FLOW

SOLE DISTRIBUTORS FOR THE UNITED STATES OF NORTH AMERICA:

D. VAN NOSTRAND COMPANY, INC.

120 Alexander Street, Princeton, N.J. (Principal office)
24 West 40th Street, New York 18, N.Y.

SOLE DISTRIBUTORS FOR CANADA:

D. VAN NOSTRAND COMPANY (CANADA), LTD.

25 Hollinger Road, Toronto 16

SOLE DISTRIBUTORS FOR THE BRITISH COMMONWEALTH EXCLUDING CANADA:

D. VAN NOSTRAND COMPANY, LTD.

358 Kensington High Street, London, W. 14

THEORY
OF TRAFFIC FLOW

PROCEEDINGS OF THE SYMPOSIUM ON
THE THEORY OF TRAFFIC FLOW,
HELD AT THE GENERAL MOTORS RESEARCH LABORATORIES,
WARREN, MICHIGAN (U.S.A.)

Edited by

ROBERT HERMAN

Department of Theoretical Physics,
Research Laboratories General Motors Corporation
Warren, Michigan (U.S.A.)

ELSEVIER PUBLISHING COMPANY

AMSTERDAM - LONDON - NEW YORK - PRINCETON

1961

Library of Congress Catalog Card Number 61 - 8865

With 122 illustrations and 20 tables

PRINTED IN THE NETHERLANDS BY N.V. DRUKKERIJ G. J. THIEME, NIJMEGEN

Preface

These proceedings are the papers presented at a Symposium on the Theory of Traffic Flow which was held at the Research Laboratories, General Motors Corporation, Warren, Michigan, on December 7 and 8, 1959. This Symposium was the third in an annual series that was started in 1957 by Dr. L. R. Hafstad, Vice President in charge of the Research Laboratories, for the purpose of providing a forum in areas of science and technology that are of special interest to the Research Laboratories as well as to the technical community at large. It has been our aim to bring together active research workers in these fields to affect cross-fertilization and to stimulate new ideas for future research activity.

The problem of traffic flow, one of Society's major problems, is an ancient one that has been of great importance whenever men or goods have been moved from one place to another. While there have been many transportation crises throughout history, we are naturally most impressed with the magnitude and frequency of modern traffic difficulties. To be sure, our engineers have done a commendable job in creating the transportation complex that has been so largely responsible for the remarkable growth and development of our Society. A number of investigators during the past few years have felt that the logical extension of research in this field would require a broader academic approach. One very important aspect of this general field is that of vehicular traffic flow. Our specific objective was to see whether it was possible to develop broad generalizations and quantitative system theories that would explain various phenomena concerning the movement of vehicles controlled by human beings. Although relatively little had been done along such lines, a sufficient groundwork had been laid to make us feel that the time was ripe for such a Symposium. In order to ensure a fresh approach, research workers from many disciplines were invited. We are more than ever convinced that significant inroads into this complex problem will be made by utilizing the findings of psychologists, physicists, engineers, physicians and many others who investigate various facets of the traffic problem, each from his own special vantage point.

The papers presented at the Symposium were limited to a relatively small number in order to allow ample time for presentation and for informal discussion which constituted a significant part of the proceedings. This arrangement was particularly appropriate because of the diversity of point of view and methodology of the various papers presented. In all, fifteen papers are published in these proceedings, of which fourteen are those that had originally been scheduled. It is very gratifying to note that the fifteenth paper, namely, that of Professor Morrell Cohen, was the result of his having been inspired by the research reported by Mr. J. G.

Wardrop. Professor Cohen's paper, which is an amplified version of his informal discussion, is the only part of such discussion that is included in this book.

The planning and execution of this Symposium was, of course, the result of the efforts of many individuals. We should like to thank various members of the Administrative Engineering Department who, as a local arrangements committee, handled many of the details necessary for the smooth functioning of such a meeting. In addition, I wish to express my appreciation to my secretary, Mrs. Margaret Hunsberger, for her devoted and competent handling of a host of problems, both expected and unexpected, that arose during the preparations and the course of the meeting. Also, I am greatly indebted to my colleagues, Dr. Denos Gazis and Mr. Richard Rothery, for giving much of their time and advice to me in connection with the technical editing of the Symposium manuscripts.

Finally, I should like to thank all of the contributors and the participants who made this Symposium a very stimulating and worthwhile meeting. In particular, I wish to thank Sir Charles Goodeve, Director, The British Iron and Steel Research Association, for his warm and enthusiastic support of our efforts during and after the meeting and for his delightful after-dinner talk.

ROBERT HERMAN

Warren, Michigan, September, 1960

Contents

The Car–Road Complex

J. B. BIDWELL

Research Laboratories, General Motors Corporation, Warren, Michigan

ABSTRACT

Transportation system problems may be broadly divided into those concerned with utilization, design, and construction. Present vehicles and highways have evolved with little formal system consideration. Emphasis has been on design and construction although recent studies have been aimed at interrelations between drivers and vehicles, including attempts at providing automatic control functions. Some type of automatic spacing control promises improved road utilization and traffic flow by increasing road capacity and safe speed. Analysis of traffic problems with human operators will require a better model of the driver than any presently available.

INTRODUCTION

This paper will not attempt to provide solutions for the many problems which have arisen as a result of the rapid expansion in private car usage. Rather, it will be concerned with some general observations concerning the nature of personal transportation problems and, more specifically, factors which influence traffic flow. Work on techniques for automatic control of vehicular traffic is also described.

The increasing severity of the traffic problem need not be substantiated by statistics, our own daily experiences are much more tangible evidence. Because of the complexity of the problem and its many interrelated factors, significant improvements may only be expected if a concerted effort is made to apply useful methods and knowledge from many disciplines.

TRANSPORTATION PROBLEMS

To provide perspective it is desirable to consider first the general transportation system problem areas and the role of traffic. Fig. 1 is an attempt at such an outline applicable to all types of transportation systems. Fundamentally, transportation implies transfer of some cargo from an origin to a destination. I have broadly divided the problems associated with accomplishing such transfer into three classes,

those related to system utilization, system design, and system construction. Each
of these may, of course, be further subdivided. Perhaps I have included somewhat
more detail in the design area because of greater familiarity with this phase. When
applied to a specific transportation system, this outline points out the problems
which must be studied. For example, when we consider the longitudinal pro-
pulsion problem for passenger cars, we must study tires and their reactions with
the road surface. Aerodynamic drag of the structure must also be considered. The
inertial properties of the vehicle and normal forces between the tire and ground

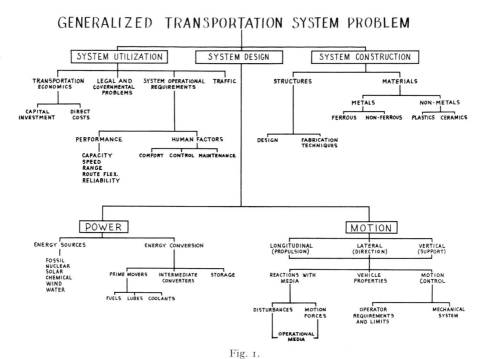

Fig. 1.

influence its forward motion. Mechanical means for regulating the motion and the
control presented to the operator must also be devised. Clearly there are close
interrelations between problem areas since the longitudinal motion performance
of the vehicle and driver will strongly influence traffic behavior which I have
classed as a system utilization problem.

Although it has been frequently observed that the vehicles and highways which
make up our present extensive personal transportation system have evolved with
little formal consideration of the entire system in the sense of modern system
analysis, still the result has been very effective in providing the majority of our
population with a means for satisfying their needs and desires in transportation.

Perhaps the principal reason for developing vehicles and roads somewhat independently is the difficulty of defining a 'mission' for a personal transportation
system and establishing system requirements.

Traffic is one of the important aspects of a transport system which influence its
utility and therefore value to users. An understanding of the effect of modifying
system components on traffic behavior is essential to the engineer.

Passenger cars are used under a wide variety of circumstances: from slow speed,
short distance, urban trips to high speed inter-city operation. They may carry

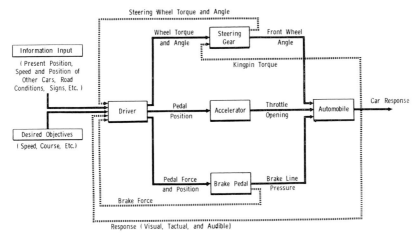

Fig. 2. Vehicle-driver system.

one person or an entire family with all of their vacation clothes and equipment.
Such basic characteristics as speed and capacity are considered by car buyers along
with comfort and appearance. Because of the great variations in both functional
requirements and buyer desires, it is difficult to assign relative values to such
diverse characteristics as capacity, comfort, range, and appearance. Unless this
can be done, synthesis of an optimum system is not possible.

CAR MOTION CONTROL

Proceeding to a somewhat smaller scale system consideration, we will look at
the problem of car control, including the driver, car and highway. It is the performance of this control system which determines the motions of aggregates of
vehicles. A simplified block diagram of the interrelations between the system
elements is shown in Fig. 2.

During recent years a good deal of attention has been given to the analysis of
vehicle response to driver and external disturbance inputs. Considerable success

has been achieved in representing this block of the system in mathematical form, and both lateral and longitudinal response to control inputs may be computed with reasonable accuracy. It is thus possible for the engineer to design vehicles to have specified response properties.

The driver provides control inputs in a conventional vehicle through the steering wheel, accelerator pedal, and brake pedal. These three inputs control the direction and speed of the vehicle. This is accomplished somewhat indirectly, however. Steering wheel displacement produces a front wheel angle which results in the desired heading change only after a period of time determined by the car speed and its response characteristics. Similarly, the accelerator regulates the opening of the throttle valve in the engine to control drive wheel torque and the brake pedal produces hydraulic pressure which results in car deceleration. The driver must manipulate these two elements to finally regulate the car speed properly.

Broadly, the driver's task consists of three functions—perception, decision, and motor response. The road defines the desired path for the driver and this, plus other information obtained by a variety of senses, serve as inputs to this block. The driver utilizes visual information about his own position and the relative location of other vehicles or stationary obstacles, senses his own car's motion kinesthetically, aurally, and visually; predicts the future driving situation taking into account natural limitations such as road friction properties and car limitations imposed by available power, braking and steering performance, and produces suitable motor responses.

Although considerable progress has been made with respect to understanding vehicle performance, this is not the case with respect to the driver, at least insofar as quantitative results are concerned. We are not, at the present time, able to express quantitatively the functional relationships between driver inputs and his response. In fact, although it is possible by properly designed experiments to derive such a functional relationship for reasonably limited inputs and with a specific control system and environment, it does not at present appear feasible to describe the driver's performance in a sufficiently complete manner so that the influence of all of the external inputs may be included. This situation prevents our prediction of his performance with completely different control systems. For example, if the driver were presented with a system which controlled yaw acceleration instead of yaw rate as in our present vehicles, or if he were given a control which produced a front wheel angle proportional to an applied torque with no displacement of the control element, we would be unable to predict his performance. It would be necessary to build the apparatus and determine performance by experiment. The integration of speed and directional control in the single stick arrangement shown in Fig. 3 is another example of a control configuration which must be evaluated by test because of our lack of ability to represent the driver and his performance in a sufficiently general and quantitative fashion. Similarly, with respect to traffic behavior of an assembly of vehicles and drivers, it is not presently possible to predict how a change in the road or in the vehicle characteristics would affect the system. We are able, of course, to investigate by means of experimental tests the influence of specific highway environment changes on driving behavior. For example, there have been several investigators who have

Fig. 3. Unicontrol.

reported on the influence of factors such as lane width, traffic density, and weather on traffic behavior. A further problem introduced in studies of driver behavior is the wide range of skill and variation in judgment and perception ability encountered in the driving population. These variations necessitate the use of statistical techniques in testing and require much larger scale experiments to assure significant results.

This situation is not wholly satisfying, for while we can by proper experiment analyze the driver control system, we are unable to synthesize an optimum system as frequently may be done with mechanical elements having accurate mathematical representations. We would like to design vehicles and roads which would have characteristics which would provide best performance when operated by the driving population, but at present cannot. The existence of this situation does not mean that progress is impossible but rather that improvements will only come about through careful experimental evaluation of proposed system modifications.

In such studies every attempt should be made to devise sufficiently accurate models of the system so that further parameter studies may be accomplished analytically. It is not even essential to obtain a complete understanding of why the system performs as it does so long as the model behaves like the real system and may be manipulated to determine how controllable variables influence performance.

AUTOMATIC VEHICLE CONTROL

The observation that driver performance is difficult to predict, coupled with the possibility of providing improved overall system efficiency, has led some investigators to consider the possibility of replacing the driver, so far as vehicle control is concerned, with some type of automatic system. The incentives for devising such a system are centered around potential improvements in safety and in traffic flow. However, the extremely complex task performed by the driver in the most general driving situation appears almost impossible to achieve using completely automatic systems, and proposals thus far advanced have been applicable only to certain limited circumstances. It is difficult to accomplish all of the sensing and logic used by the driver and this has resulted in considering automatic control primarily for application to limited access highways which involve relatively little of these functions. Even with this restriction automatic vehicle control poses many problems—technical, social and legal.

Several significant steps toward practical automatic control systems have already been made. The single stick control lever in the vehicle shown in Fig. 3 produces electrical commands which actuate electrohydraulic servos to operate steering, throttle, and brakes. These servos thus perform the usual motor function of the driver. Path error signals derived from an electromagnetic field about a wire in the road and tuned coils on the car have been used to provide steering command signals as shown schematically in Fig. 4. Such a system was publicly shown in operation at the General Motors Technical Center in February, 1958. It should be noted that a thorough understanding of the dynamic behavior of the car as well as the servo is required to produce a successful guidance system of this type.

This solution to the guidance problem illustrates a construction utilizing components both in the road and car. Alternate methods may be conceived which allocate the necessary functions in a different manner. From the standpoint of simple transition to automatic control, it would certainly be desirable to have all the necessary equipment in the car so that no modifications would be needed in the road system. Certain functions, especially perception of variables external to the car, are difficult to perform with car based equipment. This has led to complete automatic car control systems with cooperative equipment in car and road for sensing, decision, and motor functions.

SYSTEM COMPONENTS

A-ELECTRONIC CONTROL CENTER
B-VELOCITY TRANSDUCER
C-TRANSISTOR POWER SUPPLY
D-ELECTRO-HYDRAULIC CONTROL VALVE
E-GUIDANCE TRANSDUCERS
F-POSITION SERVO POSITION TRANSDUCER

1-RESERVOIR
2-HYDRAULIC FILTER
3-ACCUMULATOR
4-POWER PISTON
5-MANUAL SWITCHOVER VALVE
6-HYDRAULIC PUMP

·············· TO CONTROL CENTER

MAGNETIC FLUX LINES
ROAD GUIDANCE WIRE
ROAD CURRENT
MA·7820

Fig. 4. Automatic car guidance system.

One proposal for an automatic highway system is illustrated by the operating scale model shown in Fig. 5. Directional control of the model cars is achieved in essentially the same fashion as demonstrated in our full scale vehicles. Speed is regulated to a safe value in the following manner: the road is divided into a series of blocks each equipped with an electromagnetic detection system for establishing the presence or absence of vehicles and an antenna for transmitting a command speed for vehicles in the block. The obstacle detector consists of a primary loop carrying an alternating current and secondary loops connected in opposition so that no unbalance exists when the road is clear. A vehicle over one of the secondaries results in an unbalance due to induced eddy currents in the metal.

Safe command speeds for each block in the system are predetermined from block length and attainable deceleration rates and transmitted to vehicles by a pulse width modulated carrier. In the model shown in Fig. 5, a stopped vehicle results in a zero speed command in its own block and the block immediately behind and a half speed signal in the next block.

On a limited access road the car need only follow the path defined by the road and regulate its speed to avoid collision. Only obstacles in the same lane need be

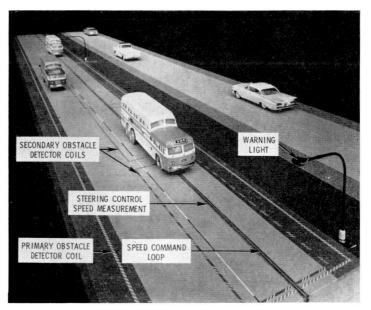

Fig. 5. Automatic highway system.

avoided. Even with such a relatively simple traffic situation, however, the difficulty of achieving adequate reliability is probably the most severe technical obstacle. Although drivers are frequently criticized for high accident rates, it would be difficult if not impossible to replace them with automatic devices of equal reliability at the present time even for the limited operation just discussed. This situation will undoubtedly change as electronic and mechanical technology advance and such controls may be anticipated in the future. Redundancy can be used to increase reliability of critical functions. In the case of the road model just described, both the primary steering signal and the speed command signal may be used to control direction. In addition further system safety is provided by the fact that loss of the speed command signal causes the vehicle to stop.

Aside from safety and convenience considerations, it is possible also to utilize automatic spacing control both to increase highway capacity and reduce transit time. Drivers presently regulate the spacing of their vehicles to accomplish safe operation (under most circumstances) when perturbations occur in spite of the lag in their response. The spacing–speed relation has been observed and found to depend a great deal on the driving environment. A typical speed–capacity curve is shown in Fig. 6. If cars could be operated bumper to bumper with no lags in the velocity response, the capacity would continuously increase with speed, as shown by the solid curve. The difference between these two curves evidently represents the

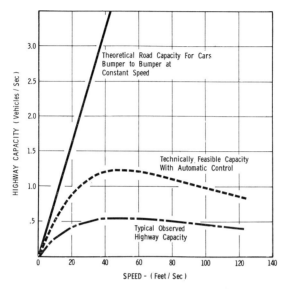

Fig. 6. Speed–capacity curve.

maximum attainable capacity improvement. Technically feasible automatic controls can about double the capacity of current roads. Such improvement potential provides a strong economic incentive for such a development. One of the most important technical problems which still remains is that of making the control self-adaptable to road properties affected by rain or other weather variables.

CONCLUSION

The foregoing discussion has served to indicate the difficulty of synthesizing control systems involving human operators to achieve optimum performance. For the near future improvements in passenger car traffic will come from analysis of factors influencing its operation and the development of manual control systems and devices to improve driver perception and response and thus system performance. Ultimately, some automatic control devices seem both feasible and desirable for limited vehicle operation.

Some Problems in Intersection Traffic Control

D. L. GERLOUGH

Ramo-Wooldridge, Canoga, Park, California

ABSTRACT

In Part I an analysis of existing traffic signal systems reveals only a minimum of feedback in even the most modern systems.

In Part II various models for links and nodes are discussed and their transfer functions presented. The dissipation of a platoon as it moves along a link appears to be analogous to the behavior of particles which are treated by wave mechanics.

INTRODUCTION

The flow of traffic on city streets is generally limited by the capacity of intersections. Where traffic control requires the installation of traffic signals, the intersection capacity is inherently reduced by the ratio of the red time to the total traffic signal cycle. Further reductions in capacity result from the time required to get a traffic stream moving once it has been stopped, by turning movement interferences, etc. By virtue of these conditions there is thus a boundary to the effectiveness of a traffic signal system. Existing traffic signal systems, however, are in many cases a long way from this boundary, and thus can be subject to substantial improvement. It is the objective of this paper to examine some of the problems underlying the control of traffic at intersections, to analyze existing intersection control systems, to identify those areas in which further theoretical study and equipment improvement are required, and to give a progress report on the attack on one of the problems. Insofar as possible, the discussion will be in terms of the terminology and methods of analysis used in connection with automatic control systems (see appendix A).

Part I

EXISTING TRAFFIC SIGNAL SYSTEMS

Although crude techniques are available for handling traffic along a thoroughfare or within a network, still a large proportion of traffic signals throughout the United States are essentially controlling traffic on an individual-intersection basis. For example, in the city of Los Angeles less than 20% of the traffic signals are connected in such a manner that they may be maintained in an appropriate time relationship to other traffic signals nearby.

References p. 23/24

Fig. 1 is a schematic plan of a typical intersection, which is included herewith for the sake of establishing nomenclature. Here the larger street is designated as street *A* and the smaller street is designated as street *B*. It is not essential that the streets be different; however, this is the most frequent type of intersection that one finds in practice.

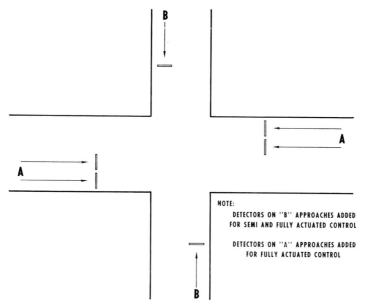

Fig. 1. General form of intersection.

FIXED-TIME CONTROL

Traffic control, when required at intersections, may be accomplished by stop signs, simple traffic signals, or very complex signal systems, together with channelization, etc. When signalization is appropriate, the type most commonly used is the fixed-time signal. A block diagram of such a control is shown in Fig. 2. Items to note here are the 'delay' blocks and the preset timing on the traffic signal controller. It is of considerable importance here that there are no feedback loops.

The motorist sees this signal control system only as it produces the delay to his travel. It should be noted that since no feedback loop is present, the control is made without regard to the delay being produced or the amount of traffic present. That is, the control system is not responsive to the needs of the traffic actually present. In truly automatic control systems, feedback is usually used to compensate for the designer's lack of knowledge as to the exact nature of the process under control. MOORE[1] states the situation very aptly when he says: "Although it is theoretically possible to make a completely open-cycle control system

References p. 23/24

as accurate as a completely closed-cycle system by knowing almost everything about the load and unalterable elements, the computer required for such a system would usually be impossibly complicated." Many automatic control systems make use of predictive information (on a feed-forward basis) for coarse setting, augmented by feed-back information for fine setting and for short-time variations.

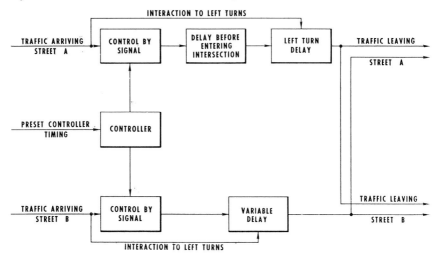

Fig. 2. Fixed-time traffic signal control system.

VEHICLE-ACTUATED CONTROL

To make traffic signals in some measure responsive to the traffic needs, traffic actuated signal controllers are occasionally installed. In the simplest of such installations, detectors or pickups are placed on the side street only, as shown in Fig. 1. Fig. 3 is the block diagram representing such a control system. It will be seen that, although this controller senses traffic arriving on the side street, this is still essentially an open-loop form of control. That is, there is no measurement of the delay or an attempt to minimize delay.

A fully-actuated traffic signal system is one in which there are pickups on all approaches, as shown in Fig. 1. Fig. 4 is the block diagram of such a control system. Again, even though pickups have been provided to measure the arriving traffic, the control is still on an open-loop basis. Fig. 5 is the block diagram representation of a more elaborate vehicle-actuated traffic signal controller which has a minimum of feedback. From this standpoint, this controller represents the most satisfactory device available today. The manner in which some agencies operate it, however, renders its value somewhat questionable for those agencies.

Fig. 6 represents the form of traffic control which would be desirable at intersections. Here the heavy lines indicate feedback paths. The open-loop data should

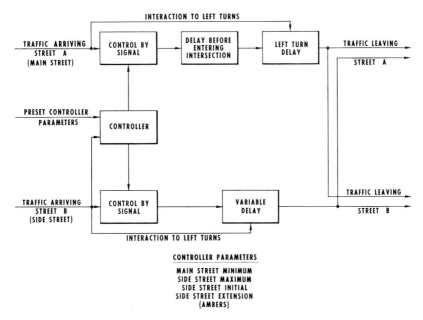

Fig. 3. Semi-actuated traffic signal control system.

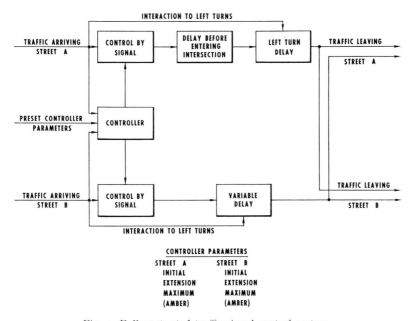

Fig. 4. Fully-actuated traffic signal control system.

include a prediction of general traffic conditions based on the time of day and on special conditions such as large gatherings of people, weather, etc. In addition, the open-loop data should include a prediction of the traffic conditions during the next signal cycle, based on measurements at contiguous intersections, on speeds between intersections, etc. The local traffic measuring unit should sense the actual arrivals at the intersection being controlled, and the delay measuring unit should provide feedback as a measure of performance.

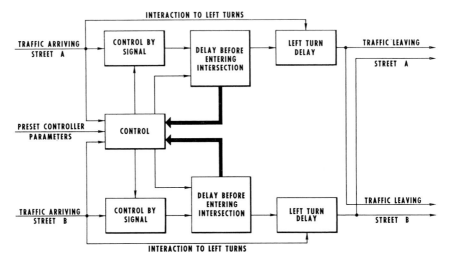

Fig. 5. Fully-actuated 'volume-density' traffic signal control system.

ARTERIALS

Considering the flow of traffic along an arterial, optimum behavior results when the traffic moves in bunches or platoons from one signal to another, arriving at the second signal just before it turns green. This type of operation is desirable for two reasons: In the first place, it is well established that an intersection will handle more traffic if the traffic arrives and passes through the intersection without stopping, thereby eliminating the dead time required to get a stream of traffic moving. In the second place, the drivers find such a system more desirable because the necessity of stopping is eliminated. In an attempt to time signals for such performance, traffic engineers frequently use graphical solutions to establish the inter-relationship between the signal timing at adjacent intersections. It has been recognized by traffic engineers that platoons tend to be dissipated as the distance between traffic signals increases. As a result, there have been established various 'rules of thumb' which lead to the installation of signals at some locations solely for the purpose of maintaining platoons.

References p. 23/24

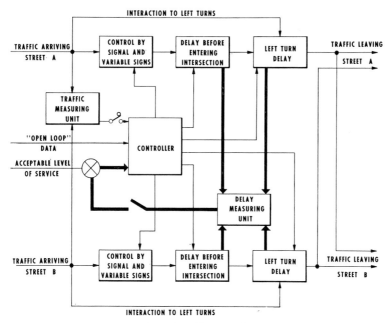

Fig. 6. 'Ideal' traffic control system.

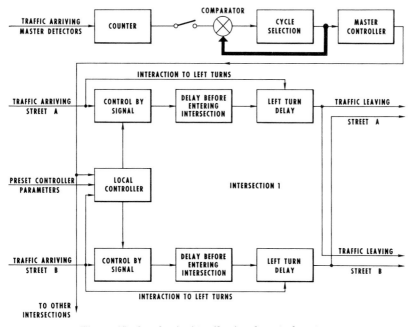

Fig. 7. 'Cycle selection' traffic signal control system.

As the volume (flow) on an arterial increases, improved performance is usually obtained by increasing the total signal cycle, and by allocating green to the side street only when there is traffic present. Fig. 7 shows the configuration of the traffic control system which is considered the most advanced in operation today. It will be noted that there is only one feedback path of rather insignificant nature, and that information is supplied to the various intersections on an essentially open-loop basis.

MEASURE OF EFFECTIVENESS

Although the above discussion has assumed total (or average) delay as a measure of performance, one of the greatest sources of concern in devising adequate traffic control systems at intersections is the definition of an appropriate measure of effectiveness. Some of the possible measures of effectiveness are:

1. Car-seconds delay produced by the traffic control system averaged over all vehicles using the intersection.
2. Percentage of cars delayed by the traffic control system.
3. Average car-seconds delay for vehicles which are delayed.
4. Average travel time through the system.

None of these measures of effectiveness may be entirely satisfactory. If one tries to optimize (*i.e.* minimize) average travel time through a system, in certain cases the optimum may result from requiring the driver to wait a rather long time at one intersection and then allowing him to proceed through a series of intersections without delay. This technique may not, however, be the one most acceptable to the driver. There are indications that the driver tension tends to increase as some higher order function of the delay at a given intersection, and that several short delays, even though of greater total duration, may be more acceptable to the driver than one long single delay.

Once an adequate measure of effectiveness has been defined, there is the problem of devising some system of sensors and method of measurement. For example, to date no adequate method has been suggested for automatically measuring the travel time through a system.

SUMMARY OF PART I

Thus far we have seen that, whereas automatic control systems ideally have closed cycles, present traffic signal systems have a minimum of feedback. There is, furthermore, some lack of agreement on the measure of effectiveness to be optimized.

Traffic signal performance is improved if traffic arrives in bunches or platoons which can pass through the intersection during the green portion of the signal cycle without being required to stop.

References p. 23/24

Part II

TRAFFIC MOVEMENT IN A STREET NETWORK

Let us now consider in greater detail some problems associated with the movement of traffic through an intersection, along the street to the next intersection, and through the next intersection.

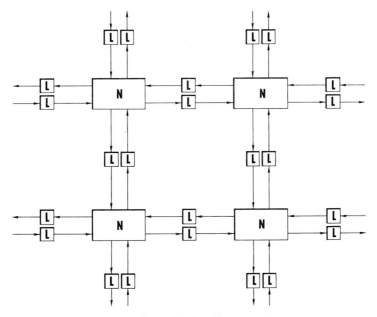

Fig. 8. Generalized traffic network.

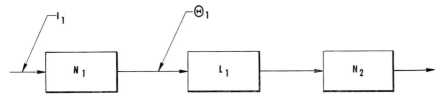

Fig. 9. Intersection–link system.

Fig. 8 shows the generalized form of a traffic network, with the intersections as nodes and the street sections as links. The various N's represent the transfer functions (see appendix A) of their respective nodes and the various L's represent the various transfer functions of their respective links. Fig. 9 is the block diagram of a system involving one lane of traffic moving through an intersection (node) N_1, traversing a street section (link) L_1, and arriving at the second intersection N_2.

TRAFFIC INPUTS

Ideally, the traffic input to N_1 would consist of a train of rectangular waves, as shown in Fig. 10. This may be stated analytically (for nomenclature see appendices)

$$I_1(s) = \frac{\exp[-st_a] - \exp[-st_b]}{s} + \frac{\exp[-s(t_a - T)] - \exp[-s(t_b - T)]}{s} +$$

$$+ \cdots + \frac{\exp[-s(t_a - nT)] - \exp[-s(t_b - nT)]}{s} =$$

$$= \frac{1}{s} \sum_{i=0}^{n} \{\exp[-s(t_a - iT)] - \exp[-s(t_b - iT)]\}$$

In such cases, the transfer function of the intersection, N_1, need be only such that it will match the input wave and pass it without distortion. Unfortunately, most realistic intersections are more complex.

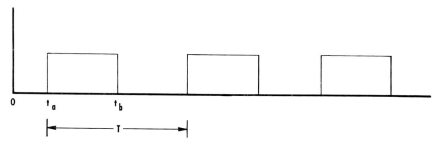

Fig. 10. Ideal waves of traffic.

There is considerable difficulty, however, in finding an adequate statistical representation for traffic in the intersection environment. When traffic is some distance (usually greater than $\frac{1}{2}$ mile) past a signal (or stop sign) it behaves in a random manner. In this case, however, 'random' does not necessarily imply 'Poisson'. When traffic becomes the least bit heavy, departures from the Poisson distribution occur. A recent paper by HAIGHT[2] discusses a family of distributions which HAIGHT refers to as the "generalized Poisson distribution". Recent fragmentary unpublished work by HAIGHT and by the writer suggests that there may be a functional relationship between the generalized Poisson parameter and the density of traffic.

Associated with the adequate representation of traffic distribution is the problem of detecting a change in the level of traffic in order to make appropriate changes in the control system (*e.g.* signal timing settings). If control is to be performed on the basis of density, it may be necessary to examine transient conditions if density is near the critical value. If control is to be based on volume only, it appears that smoothing of data over suitably short sampling periods may be acceptable. Fig. 11 is a plot showing the build-up of traffic on a particular facility. The points represent 6-minute samples. The heavy envelope is drawn to permit analysis. This envelope has been analyzed by Fourier transform methods[3]. A continuous plot of the frequency spectrum is very tedious. For the present purposes, however, only the

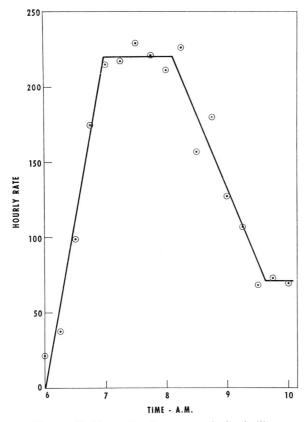

Fig. 11. Build-up of traffic on a particular facility.

maxima are required. It is very easy to determine the envelope of these frequency maxima. Such an envelope has been normalized in terms of the amplitude for one cycle per hour and is plotted in Fig. 12. It becomes a matter of judgement to select the frequency beyond which amplitudes are negligible. In the present example it appears that frequencies beyond 10 cycles per hour may be neglected.

It is general rule of sampled-data control systems that the sampling frequency must be at least twice the highest frequency present[4,5]. Thus, sampling of the traffic in the situation used as an example should occur no less than 20 times per hour; sampling rates 25 or 30 times per hour would be preferable.

Of course, not all traffic facilities will have as steep a build-up as that shown in Fig. 11. Before a sampling control system is installed, the sampling period should be determined by a frequency spectrum analysis.

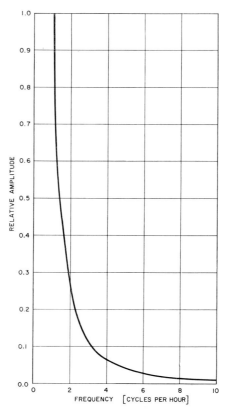

Fig. 12. Envelope of frequency spectrum of traffic build-up.

LINK BEHAVIOR

If the link between nodes could be considered only as a delay with no 'smearing', the transfer function of the link would be

$$L(s) = \exp\left[-\ (x/v)s\right]$$

where x = distance between nodes, v = speed.

The link, however, appears to behave more like a noninductive line having uniformly distributed resistance and capacity. Using a link of this form leads to a transfer function of the following type

$$L(s) = \exp\left[-\,kx\,(s)^{\frac{1}{2}}\right]$$

where x = distance, k = a constant which characterizes the particular link and includes speed.

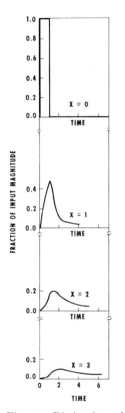

Fig. 13. Dissipation of pulse in a noninductive line.

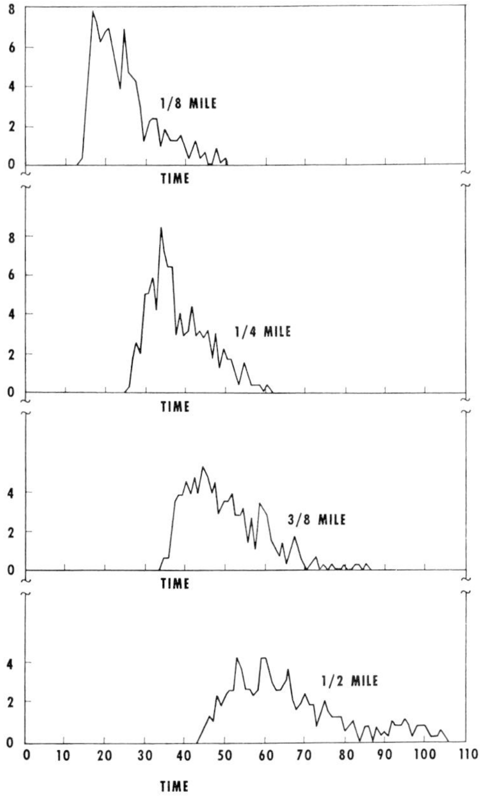

Fig. 14. Dissipation of traffic platoon.

This transfer function is derived in appendix B. When a link having this transfer function receives an input of rectangular form as shown in Fig. 10, of height I and where $t_a = a$ and $t_b = b$, the transform of the input is

$$I(s) = \frac{\exp\left[-\,as\right]}{s} - \frac{\exp\left[-\,bs\right]}{s}$$

References p. 23/24

which, when multiplied by the transfer function gives

$$\Theta(s) = \frac{\exp\left[-kx(s)^{\frac{1}{2}}\right]}{s} (\exp\left[-as\right] - \exp\left[-bs\right])$$

The inverse transformation gives

$$\theta(t) = \operatorname{cerf} \frac{kx}{2\ (t-a)^{\frac{1}{2}}} - \operatorname{cerf} \frac{kx}{2\ (t-b)^{\frac{1}{2}}}$$

Fig. 13 illustrates the dissipation of a pulse through such a line. In this figure $a = 0,\ b = 1$.

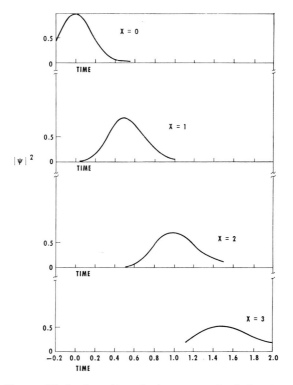

Fig. 15. Dissipation of impulse in wave-mechanical system.

Fig. 14 shows the dissipation of a traffic platoon. A comparison of Fig. 13 and Fig. 14 suggests that the analogy between a traffic link and a noninductive line may be a useful first approximation.

The foregoing treatments of links have dealt with traffic as though it were continuous in nature. Traffic, on the other hand, is discrete in nature even though

References p. 23/24

it exhibits wave-like properties. The field of wave mechanics has been developed by physicists to treat certain systems which have these properties. It would appear therefore, that the methods of analysis used in wave mechanics might be applicable to the behavior of traffic along a link.

Using a wave mechanics approach, the following expression has been obtained (derivation in appendix C)

$$\theta = \frac{A^2\pi}{(\sigma^2 + \pi^2\alpha_2^2 t^2)^{\frac{1}{2}}} \exp\left[\frac{-2\pi^2\sigma(x - \alpha_1 t)^2}{\sigma^2 + \pi^2\alpha_2^2 t^2}\right]$$

where A and σ are constants, $\alpha_1 =$ initial group velocity and $\alpha_2 =$ initial group acceleration.

When A, σ, α_1, α_2 are assigned values, θ may be plotted as a function of t with x as a parameter. Fig. 15 is such a plot, with $A^2 = (1/\pi)$, $\sigma = 1$, $\alpha_1 = 1$, $\alpha_2 = (1/\pi)$. It will be noted that as x increases the distributions become wider and flatter, with "tails" extending to the right. On comparison of Fig. 15 with Fig. 14 it appears that with appropriate values of constants Fig. 15 might reasonably serve as a smoothed version of Fig. 14. Inasmuch as Fig. 15 represents the probability that a given particle reaches a given distance as a function of time, analysis of the type used serves to tie together deterministic and probabilistic considerations.

SUMMARY

It has been pointed out that existing traffic control systems at intersections are largely open-loop in nature. To achieve the greatest efficiency, techniques must be developed for measuring and controlling some figure of merit on a closed-loop basis. Furthermore, techniques must be developed which will permit the formation of traffic into platoons to more effectively use the traffic signal cycle.

Considerable work is needed on distributions for representing traffic and for the transfer function of both intersections and street sections.

Some interim solutions to platoon problems, that is, mathematical representations of platoons, have been proposed.

REFERENCES

1 J. R. MOORE, Combination Open-Cycle Closed-Cycle Systems, *Proc. Inst. Radio Engrs.*, *39* (1951) 1421–1432.
2 F. A. HAIGHT, The Generalized Poisson Distribution, *Ann. Inst. Statist. Math. (Tokyo)*, *11* (1959) 101–105.
3 S. GOLDMAN, *Frequency Analysis, Modulation and Noise*, McGraw-Hill, New York, 1948.
4 J. G. TRUXAL, *Automatic Feedback Control System Synthesis*, McGraw-Hill, New York, 1955, p. 505.
5 J. R. RAGAZZINI AND G. F. FRANKLIN, *Sampled-Data Control Systems*, McGraw-Hill, New York, 1958, p. 17.

6 M. F. GARDNER AND J. L. BARNES, *Transients in Linear Systems*, Wiley, New York, 1943, vol. I.

7 L. A. PIPES, *Applied Mathematics for Engineers and Physicists*, McGraw-Hill, New York, 1958.

8 E. M. GRABBE, S. RAMO, AND D. E. WOOLDRIDGE, *Handbook of Automation, Computation, and Control*, Wiley, New York, 1958, vol. I, pp. 20–61 to 20–66.

9 J. R. CARSON, *Electric Circuit Theory*, McGraw-Hill, New York, 1926, p. 86.

10 N. F. MOTT, *Elements of Wave Mechanics*, Cambridge Univ. Press, Cambridge, 1952.

11 J. FRENKEL, *Wave Mechanics, Elementary Theory*, Dover Publ., New York, 2nd ed., 1950.

12 N. F. MOTT AND I. N. SNEDDON, *Wave Mechanics and Its Applications*, Oxford Univ. Press, Oxford, 1948.

13 C. A. COULSON, *Waves*, Oliver and Boyd, London, 1949, p. 136.

Appendix A

NOTES ON SOME TECHNIQUES USED IN CONTROL SYSTEM ANALYSIS

Transfer Function

If $A(s)$ is the Laplace transform of the input to a system and $B(s)$ is the Laplace transform of the output, then $G(s)$ is a property of the system such that

$$B(s) = A(s)\, G(s) \qquad \text{or} \qquad G(s) = B(s)/A(s)$$

where $G(s)$ is called the transfer function of the system.

Laplace Transform

The Laplace transform is a technique by which differential equations can be transformed into algebraic equations. The transform is very useful in analysis of probability distributions. There are many standard texts available [6,7]. In one notation the defining equation is

$$\mathscr{L}\, f(t) = F(s) = \int_0^\infty f(t) \exp\left[-st\right] dt$$

Block Diagrams

The block diagram is widely used as a technique for analysis of certain aspects of control systems. A block is usually used to represent some function performed on an input which results in an output which differs from the input. Drawing a block diagram causes the engineer to think clearly so as to completely represent all facets of the system. Frequently, aspects of a system which are not obvious appear as the block diagram is drawn. Frequently the function performed by each block is represented as a transfer function. Once a block diagram has been drawn, it can be manipulated by certain standard rules [8] to obtain a diagram having a markedly different configuration but representing the same over-all operation.

Appendix B

TRANSFER FUNCTION OF A NONINDUCTIVE LINE

According to CARSON[9] the output of a noninductive transmission line, given a unit step input voltage, is

$$V_{out} = 1 - \frac{2}{\pi^{\frac{1}{2}}} \int_0^z \exp\left[-u^2\right] du = \text{cerf}(z)$$

where

$$z = \frac{x(RC)^{\frac{1}{2}}}{2 \, t^{\frac{1}{2}}}$$

According to PIPES[7], page 647, but with conversion to the notation of GARDNER AND BARNES[16] we have

$$\mathscr{L}\left[\text{cerf}\left(\frac{a}{2 \, t^{\frac{1}{2}}}\right)\right] = \frac{1}{s} \exp\left[-as^{\frac{1}{2}}\right]$$

where

$$a = x \, (RC)^{\frac{1}{2}}$$

But, the transform of a unit step is $1/s$, and, by definition of the transfer function given earlier

$$G(s) = \frac{\mathscr{L}V_{out}}{\mathscr{L}V_{in}}$$

Therefore, the transfer function of a noninductive line is

$$G(s) = \exp\left[-as^{\frac{1}{2}}\right]$$

where

$$a = x(RC)^{\frac{1}{2}} = K_1 \, K_2 \, x = K_3 \, x$$

The quantities K_1 and K_2, or their product, K_3, are constants which describe the line.

Appendix C

WAVE MECHANICS

DE BROGLIE[10,11] has shown that the behavior of certain material particles can be treated as wave motion. This treatment is generally probabilistic in nature. The movement of vehicles on streets and highways is a phenomenon of discrete particles and one which is frequently approached from the probabilistic viewpoint. Some of the concepts of wave mechanics appear to be suitable in the treatment of certain traffic problems.

Let p = momentum of particles, V = velocity of particles, m = mass of particles, h = a constant (in the usual case: Planck's constant), k = wave number = number of waves per unit length, v = frequency of oscillation, V_g = group velocity.

Then $p = mV$, p = hk (may be regarded as the result of experiments[12]) $v = Vk$ and V_g = dv/dk.

But the particles travel with group velocity. Therefore

$$V = V_g = \frac{dv}{dk} = \frac{hk}{m}$$

$$\int dv = \int \frac{h}{m} k \, dk$$

and

$$v = \frac{h}{m} \frac{k^2}{2} + C_1 \text{ and (where } C_1 \text{ is an arbitrary constant)}$$

$$hv = \tfrac{1}{2} mV^2 + hC_1$$

Waves moving along the x axis may be expressed by

$$\psi = a \exp [2\pi j (\pm kx - vt)]$$

where ψ is the degree of disturbance at a given place and time, $|\psi| = N^{\frac{1}{2}}$
        N is the density of particles (per unit length) and j $= (-1)^{\frac{1}{2}}$.

The parameter a will in general be a function of k in accordance with some distribution. Assume a distribution of the following form

$$a = A \exp [-\sigma (k - k_0)^2] \text{ per unit range of } k$$

Then

$$\psi(x,t) = \int_{-\infty}^{+\infty} A \exp [-\sigma (k - k_0)^2 + 2\pi j (kx - vt)] \, dk$$

Before this integration can be carried out it is necessary to have v as a function of $(k - k_0)$. Taylor's series can be used for this purpose

$$v = \alpha_0 + \alpha_1 (k - k_0) + \frac{\alpha_2 (k - k_0)^2}{2!} + \ldots$$

with

$$\alpha_0 = v_0 \qquad \alpha_1 = \left[\frac{dv}{dk}\right]_0 = \left[V_g\right]_0 \qquad \alpha_2 = \left[\frac{d^2v}{dk^2}\right]_0 = \left[a_g\right]_0$$

By truncating the series at this point, the degree of disturbance becomes

$$\psi(x,t) = \int_{-\infty}^{+\infty} A \exp\left[-\sigma(k-k_0)^2 + 2\pi j\left\{kx - t\left[v_0 + \alpha_1(k-k_0) + \tfrac{1}{2}\alpha_2(k-k_0)^2\right]\right\}\right] dk$$

$$= A \int_{-\infty}^{+\infty} \exp[-2\pi j v_0 t + 2\pi j k_0 x + 2\pi j(x-\alpha_1 t)(k-k_0) + (k-k_0)^2(-\pi j\alpha_2 t - \sigma)] dk$$

$$= A \exp[2\pi j(k_0 x - v_0 t)]\left(\frac{\pi}{\sigma + \pi j\alpha_2 t}\right)^{\frac{1}{2}} \exp\left[\frac{-\pi^2(x-\alpha_1 t)^2}{\sigma + \pi j\alpha_2 t}\right]$$

For method of integration see COULSON[13]

$$\psi(x,t) = A\left(\frac{\pi}{\sigma + \pi j\alpha_2 t}\right)^{\frac{1}{2}} \exp\left[\frac{-\pi^2\sigma(x-\alpha_1 t)^2}{\sigma^2 + \pi^2\alpha_2^2 t^2} + 2\pi j(k_0 x - v_0 t + \gamma)\right]$$

$$\psi(x,t) = \frac{A\pi^{\frac{1}{2}}}{(\sigma^2 + \pi^2\alpha_2^2 t^2)^{\frac{1}{4}}} \exp\left[\frac{-\pi^2\sigma(x-\alpha_1 t)^2}{\sigma^2 + \pi^2\alpha_2^2 t^2} + 2\pi j(k_0 x - v_0 t + \gamma) - \tfrac{1}{2}j\beta\right]$$

where

$$\gamma = \frac{\pi^2\alpha_2 t}{2(\sigma^2 + \pi^2\alpha_2^2 t^2)} \qquad\qquad \beta = \tan^{-1}\frac{\pi\alpha_2 t}{\sigma}$$

and finally

$$\theta = |\psi|^2 = \frac{A^2\pi}{(\sigma^2 + \pi^2\alpha_2^2 t^2)^{\frac{1}{2}}} \exp\left[\frac{-2\pi^2\sigma(x-\alpha_1 t)^2}{\sigma^2 + \pi^2\alpha_2^2 t^2}\right]$$

To establish the analogy between traffic and wave mechanics it will be necessary to identify the various parameters in the traffic situation. Clearly the group velocity, V_g, will be the speed of the vehicles. This will, of course, follow some statistical distribution. The quantity m in the formulas requires some thought. It is fairly obvious that m is not the masses of the vehicles. It may be some property which will be termed 'inertia' and which represents the sluggishness of the traffic to change velocity. The quantity h is presumably some constant. Once V_g, m, and h have been established, the other quantities v, k, and p can be determined by the relationships given.

Traffic Flow with Pre-Signals and the Signal Funnel

W. VON STEIN

Traffic Department, City of Düsseldorf, Germany

ABSTRACT

It is shown by means of numerous examples how traffic may be regulated in a more elastic and diversified manner, using two as yet little known types of signals: pre-signals and speed signals.

Pre-signals serve especially in guiding minor side traffic streams into the queuing area before intersections. At the same time they can help to avoid loss of starting time.

Pre-signals permit the formation of so-called time islands for streetcar stops where there is no room for loading islands. Existing loading islands may be sealed off by pre-signals, which facilitates the change of passengers and gives them more safety.

The augmentation of signal language by adding 'quicker' and 'slower' to 'stop' and 'go' by means of the speed signal allows the formation of signal funnels. They considerably improve traffic flow before and between intersections and also considerably improve safety on high speed roads.

Signal funnels constitute a suitable coordination of signal systems with differing cycles. Also individual traffic control is possible by using signal funnels actuated by vehicles or pedestrians in such a fashion that vehicles do not have to stop in the main direction.

The abolition of unnecessary stops constitutes an economic equivalent for the installation costs of signal funnels.

TRAFFIC FLOW IN SIGNAL SYSTEMS PRESENTLY USED

Under favorable conditions, it is possible in coordinated signal systems to widely avoid the stopping of vehicle groups before the second and the following intersections. The loss of starting time is diminished provided the next intersection is not blocked by stopped vehicles which have turned in from other streets.

Up to now it was unavoidable that more than half of the vehicles moving in the through direction had to stop at the first intersection of a green wave. As compared to the 'flying start', the start from the stop-line results in a loss of 3 to 4 s even for well accelerating vehicles. If the time–space relation at the first intersection cannot be extended by 4 s then, at the following intersections, it means

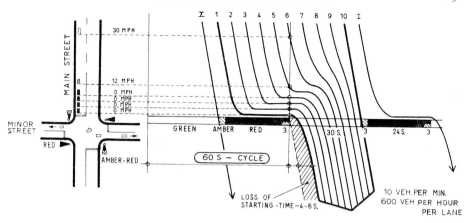

Fig. 1(a). Time–space diagram of a normal crossing with start on the stop-line.

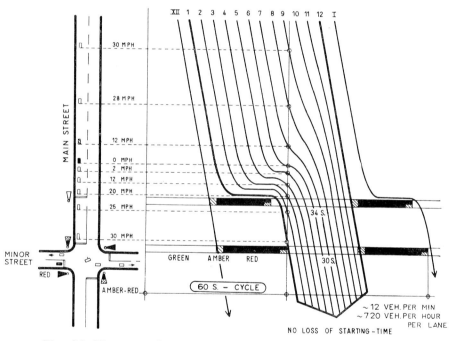

Fig. 1(b). Time–space diagram of a crossing with pre-signal and flying start.

a 14 to 16% loss of efficiency due to starting, assuming the usual green phases of 25 to 28 s.

The time–space diagram of a single intersection (Fig. 1) distinctly shows that of an average of 10 oncoming vehicles per cycle and per lane, 5 to 6 come to a stop. Only 4 to 5 proceed freely. Here it is easily discernible how the loss of starting time of the stopped vehicles delays the following free moving vehicles. At a free period of 30 s only 10 vehicles per unimpeded lane can pass after starting, instead of 12 possible vehicles at flying start.

THE PRE-SIGNAL AS A MEANS OF AVOIDING LOSS OF STARTING TIME

A pre-signal located before the intersection signal at the normal accelerating distance of 100 ft gives a lead of 5.5 s on the basis of a maximum accelerating time of 6 ft/s² at 25 mph. The pre-signal facilitates a flying start for all vehicles and widely avoids loss of starting time. In this case the loss of starting time does not occur at or after the intersection but rather ahead of it. Therefore it does not use up any green time (Fig. 1).

Heavier vehicles, such as trucks and buses, which develop smaller accelerations, arrive at the intersection after the green has been on for some seconds. In order to avoid this time loss it has been tried to let the pre-signal operate on the basis of an average acceleration of 2.6 to 3.3 ft/s². In fact, remarkable gains of time were observed when there were heavy vehicles at the front of a group. However, when passenger cars were in front, usually accelerating too fast (5 to 6.6 ft/s²), they came to the intersection when it was still showing the red light. They had to stop, thus maintaining the full loss of starting time. Consequently such pre-signals must be adjusted to an acceleration of 6.6 ft/s². In this case heavy vehicles regain only part of the loss of starting time.

The 4 s gained by the use of pre-signals and the resulting flying start seem to be insignificant, but if the green phases of the cross streets are only 12 to 16 s, this adds up to 25 to 30%.

When the traffic density approaches saturation, it may be important to avoid any loss of starting time.

THE 'DYNAMIC LOADING ISLAND' OR 'TIME ISLAND'

In narrower streets with streetcar traffic there is often no room for loading islands. When the streetcar stop is located in front of a signalized intersection, passengers alighting from and boarding the streetcar are often annoyed by lines of vehicles and even passing vehicles. Sometimes the streetcar arrives just at the beginning of the green phase and stops, thus impeding the traffic for as long as an entire light cycle. The efficiency of traffic flow decreases heavily when the street car traffic is heavy.

By erecting a pre-signal ahead of the streetcar stop a so-called 'time island' can be formed (Fig. 2). This pre-signal allows the streetcar to pass only when the last of the nonrailbound traffic flows off at the green phase. The red portion of the signal cycle can then be used for the unimpeded movement of passengers, because the other vehicles have to stop at the pre-signal showing red. The whole length of street between the streetcar and the curb remains free of vehicles. Soon after the change of passengers is accomplished, the pre-signal allows traffic to proceed. The traffic arrives at the intersection with a flying start and crosses the intersection together with the streetcar. Streetcars, passengers, and other vehicles have therefore more time and safety. This system was developed in Düsseldorf and has already been widely adopted in Germany.

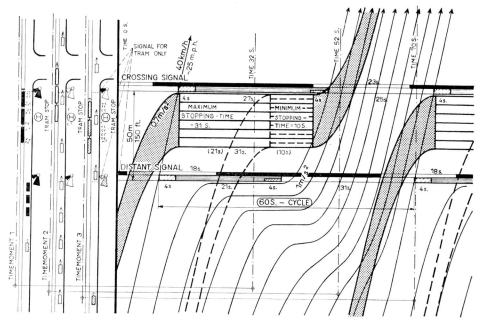

Fig. 2. Tram-stop with pre-signal (dynamic loading island). Note that the terms 'pre-signal' and 'distant signal' are used interchangeably.

Of course, a separate signal for the streetcars must be combined with the pre-signal so that the streetcars do not block access to the intersection at green. Sufficient time must be provided for the change of passengers at the stop. Marking of the pavement must ensure that the rails always remain free of vehicles in the stopping area (Figs. 2 and 3).

On account of the great advantages for the passengers, pre-signals are also often set up in front of platform-type loading islands, so that the passengers can leave the islands quickly and safely without restrictions (Figs. 4, 5 and 6). This does not impede nonrailbound traffic. On the contrary, it prevents starting time losses.

Fig. 3(a).
Time island in
Düsseldorf.

Fig. (3b).
Time island in
Düsseldorf.

Fig. 3(c).
Time island in
Düsseldorf.

Fig. 3(d).
Time island in
Düsseldorf.

Fig. 4.
Time island with
platform-type
loading island.

Fig. 5.
Time island with
platform-type
loading island.

Fig. 6.
Fig. 6. Time island with platform-type loading island.

OTHER APPLICATIONS OF PRE-SIGNALS

The application of pre-signals for 'time islands' is only a part of their possibilities. If there is an important influx of traffic within 150 ft of an intersection, this spot can very well be serviced by a pre-signal, without interrupting the main traffic flow. On the contrary, an increase of efficiency is even possible provided there is no constant influx of vehicles from this approach.

We now discuss some practical examples where the difficulties could only be solved by using a pre-signal.

A fire station has 4 big doors for simultaneous exit in the queue space before a signalized intersection. It was known from another city that a special signal before such an exit is easily overlooked, if it is either normally dark or constantly green. The drivers who are acquainted with this situation do not pay any attention to it on account of the rare changes of the signal. In Düsseldorf a pre-signal was put up on a whip pole just in front of such an exit. This signal changes with the rhythm of the main signal but slightly ahead of it in time. By this signal the queuing space has simply been moved away from the exits, pulling back the stop line by 150 ft. Furthermore, the signal has been connected to the fire alarm, so that, in case of an alarm, all access signals to the intersection including the pre-signal turn red and stay red for two full 60-s cycles. (Only the main signal behind the pre-signal shows green.) In this time the fire engines can leave the station and be on their way. This kind of operation has fully proved its usefulness for several years.

In another case, a garage for 370 cars was going to be built near the corner of an important intersection. As the authorities expected traffic jams in front of the intersection, they would have denied the building permit. By putting up a pre-signal 240 feet from the intersection, it was not only possible to regulate the exit from the garage satisfactorily, but also to make safe a streetcar stop which otherwise would have been within the queuing space.

A great industrial plant situated at another important intersection had the entrance of its parking lot about 100 ft from the intersection. After signals had been installed this entrance could only be reached by a right turn from the main street. Vehicles coming from the cross street or from the opposite direction on the main street either faced the main oncoming traffic or a queue of stopping cars. A pre-signal combined with a speed signal to form a traffic funnel now regulates the access to the parking lot permitting the left turn without difficulties.

Such cases are not rare at all. Heretofore they had mostly been solved by denying building permits or by expensive rebuilding.

If an intersection consists of 5 streets, this situation can also easily be handled by the use of pre-signals. The fifth street is merged before the intersection with the main thoroughfare in such a way that sufficient queuing space remains between this junction and the intersection itself. In the main thoroughfare a pre-signal is placed before the junction of the fifth street. When the presignal shows red, traffic from the fifth street can flow in and has to stop again before the main signal. When both signals turn green simultaneously, traffic from this fifth direction can be included in the normal cycle without special phases and without loss of time.

THE SIGNAL FUNNEL FOR CONSTANT TRAFFIC FLOW

From the simple pre-signal, intended to avoid loss of starting time, the so-called signal funnel or 'Düsseldorf funnel' was developed. It had the aim of regulating traffic already ahead of intersections and keeping it in constant flow. The funneling begins at the location where the 'speed lines', *i.e.* trajectories of the last quickest vehicle of a cycle meet the lines of the first slowest vehicle of the next cycle, on the time–space diagram. When speed is limited to a maximum of 30 mph in urban traffic, then the upper limit is fixed. The slowest admissible speed can be fixed at 20 mph. These two lines meet at about 2800 ft before the intersection. The intersection point on the time–space diagram denotes the point from where the end of the green phase may still be attained by traveling at 30 mph, or the beginning of the following green phase by traveling at 20 mph. At this point the so-called speed signal is to be placed. This signal is geared to show 3 different speeds, for instance 20, 25, and 30 mph. By observing this signal all vehicles would reach the intersection during the green period. A tryout at an intersection with a very short red period and only 1100 ft distance between the speed signal and the intersection (at Viersen, Rhineland) resulted in stops at the intersection which could not be avoided fully. At any rate the percentage of vehicles able to pass freely, *i.e.* without stopping, increased from 55% without a speed signal to 70 to 77% with a speed signal.

THE FIRST SIGNAL FUNNEL

The first signal funnel was installed in 1954 near the intersection of Witzelstrasse–Auf'm Hennekamp on the last 500 meters ($\sim$ 1600 ft) of the Witzelstrasse in Düsseldorf. The funnel consisted of 6 movable pre-signals, all shaped like the signal at the intersection. The signal at 1600 ft distance from the intersection showed green almost all the time with a short red period of 6 s. A slight retardation to about 19 mph was sufficient for the traffic to obtain the next green light. Every vehicle proceeding at the same speed reached the next signal at the beginning of the green phase. For the last vehicles the green wave was attained at 34 mph. When traffic was not too dense, an efficiency rate of 90 to 95% of all vehicles was reached. Only during peak hours were stops observed within the funnel. Without the signal funnel the percentage of freely passing vehicles was 45%. Pre-signals installed only at the right hand side of the roadway forced drivers of overtaking vehicles to brake sharply because they suddenly saw a red light in the front of them. In some cases other vehicles hit the braking vehicles. Later the red lights were replaced by flashing amber lights. As a consequence the accidents stopped, but the efficiency of traffic flow decreased considerably. Depending on the traffic density the efficiency was between 75 and 85%. At darkness, however, traffic flow was always improved, because the signals attracted the attention of the drivers better. But the drivers soon wanted to know the speed at which they would have to drive at the beginning and at the end of a green period. Thereafter the first speed signal was installed ahead of the other signals, showing at first 30 km/h, then 40 km/h and then 50 km/h in white figures indicating the correct speed. As a consequence, the percentage of the traffic flow that did not have to stop increased somewhat.

The first permanent funnel on Kruppstrasse toward Oberbilker Allee consisted of a speed signal and 2 pre-signals with red, amber and green lights, coupled with signals attached to whip poles and thus hanging over the roadway to ensure better visibility for overtaking vehicles. This installation at once proved to be no cause of accidents and allowed a traffic flow of 90 to 97% of freely passing vehicles even though at the intersection itself the green time was only 47% of the total light cycle.

Fig. 7(a).
Speed signals.

Fig. 7(b).
Speed signals.

Fig. 8.
Speed signals

Fig. 9.
Speed signals.

THE FORM OF THE SPEED SIGNAL

The first speed signal was constructed from a normal three-lens light signal using colorless dispersing lenses 8 inches in diameter. The upper lens showed the figure 30, the middle one 40, and the lower one 50, in white light on a black background. For stangers unacquainted with this installation the figures were only discernible at a distance of 150 to 250 ft. However, drivers acquainted with the signals adapted themselves to the position of the lights. The signal was posted at the right hand side only and often was covered by large vehicles (Fig. 7a).

When another funnel was installed later, 2 signals were attached to each other in such a way that the figures 4, 5, and 6 were put on the first signal and the figure 0 on the second signal. Thus visibility was obtained at a distance of 450 ft and more (Fig. 7b).

Close to the right side of the signal a metal sign was placed, showing 'km/h' next to the luminous figures, as well as the heading 'For 500 meters now'.

Reducing speed on streets with a maximum speed of more than 30 mph constituted a threat to traffic safety. Therefore, all drivers, also when overtaking, had to see the signals easily. This necessitated not only the installation of the enlarged signal on the right hand side of the roadway but also of another signal of the same size above the roadway. Thus all vehicles using the left lane could see it (Fig. 8).

At present the speed signal is not yet an official traffic sign according to the traffic codes. It is only a recommendation. For this reason it has been tried to provide the signal with a circular disk to give it a look similar to the international traffic sign for maximum speeds. This sign is shown in Fig. 9. It is not yet clear, however, whether the courts will acknowledge this mixture of a signal and a traffic sign. There are indeed traffic signs in use nowadays which change their meaning during certain hours daily. But changes within a few seconds constitute a novelty for traffic law and jurisdiction. At any rate these speed signals, the use of which daily gains in importance everywhere, should be embodied in the traffic codes. They are becoming a decisive safety factor before intersections, especially on high-speed roads.

THE STAGGERED SIGNAL FUNNEL

Continued observations of traffic flow in signal funnels showed that for straight speed lines (Fig. 10) the first vehicles of a group tend to arrive too early at the signals, being forced to stop there. On the other hand the last vehicles often could not maintain the recommended maximum speed during dense traffic because they were impeded by slower vehicles. In order to eliminate these difficulties the staggered signal funnel was developed. This innovation (Fig. 11) operates so that the first vehicles of a group at the first pre-signal are retarded a little more than would

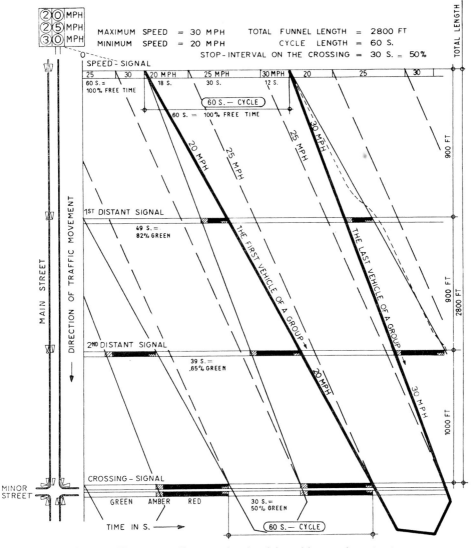

Fig. 10. Time–space diagram of a signal funnel in an urban street.

have been necessary for vehicles correctly maintaining the recommended speed. A vehicle arriving too early had to slow down somewhat, but then it certainly did not arrive at the intersection before the beginning of the green phase. The last vehicles again were forced to accelerate a little more before the pre-signal to enable them to reach the end of the green phase at the intersection. This is true even if the density of traffic would force a speed decrease when approaching the end of the funnel. An expert driver will know when reaching a pre-signal at the

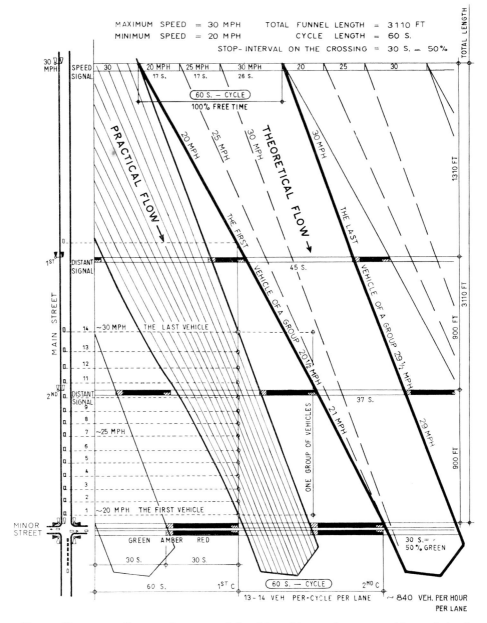

Fig. 11. Time–space diagram of a staggered signal funnel in an urban street with speed signal and two pre-distant signals.

end of a group just at changing time that he will safely pass the intersection if he maintains his speed. If one looks at the speed lines on the time–space diagram of such a funnel, they appear to be somewhat bent at the pre-signals. The speeds are staggered from signal to signal. The narrowing of the funnel is no longer constant. It decreases a little toward the intersection. With the staggered funnel the condensation of a group of vehicles is achieved predominantly in the first part of the funnel. In the second and third parts of the funnel there is only a slight additional condensation. Thereby an essential success in traffic flow is attained.

Further refinement in traffic signals was embodied in the staggered signal funnel. Comparing Fig. 10 and Fig. 11, one finds that the first vehicle reading 25 mph at the signal will reach the intersection exactly at the beginning of the green phase, if not impeded. Thus it is forced to overtake the vehicles of the preceding group. The last vehicle getting 25 mph arrived at the intersection exactly at the end of the green phase. It was overtaken by the rest of the group at 30 mph. After observing that these overtakings caused no difficulties we advanced the change from 25 mph to 30 mph so that the first vehicles at 30 mph also arrived at the intersection shortly after the beginning of the green phase (see Fig. 11). This method is now applied exclusively.

LENGTH OF THE FUNNEL AND NUMBER OF SIGNALS

The length of the funnel is a function of the fixed maximum and minimum speeds and also of the length of the red period at the intersection. The greater the difference between maximum and minimum speeds, the shorter the funnel. Longer red periods at the intersections and staggering of pre-signals lengthen the funnel. Experience has shown that the best results are obtained at speeds between 20 and 30 mph in urban areas and between 25 and 40 mph in other surroundings. A speed higher than 45 mph cannot be permitted at signalized intersections for safety reasons. The average driver does not have the necessary experience regarding braking from such high speeds. In urban areas at minimum speeds of 20 mph and maximum speeds of 30 mph and at red light periods of 20 s the shortest funnel will have a length of 1900 ft. Each additional second of red or amber time lengthens the funnel by 100 ft. The staggering of speeds means 200 ft more. High-speed roads outside of urban areas can be operated at minimum speeds of 25 mph and maximum speeds of 40 mph. The shortest funnel will then have a length of 2000 ft with a red period of 20 s. With staggering of speeds the length will be 2200 ft. It does not seem advisable to keep the funnel short. If it is longer, greater efficiency of traffic flow can be expected.

The simplest fairly efficient funnel consists of a speed signal posted at the opening of the funnel, a pre-signal at its middle, and the intersection signal itself.

Funnels of this kind have proved their efficiency in several locations. With longer funnels it is recommended to install two pre-signals because dispersing effects make themselves felt if the distance between signals is greater than 1000 ft.

FUNNELS WITH TWO SPEED SIGNALS

Drivers who are not yet acquainted with the principle of the signal funnel often do not pay sufficient attention to the first speed signal when they are driving at a considerably higher speed. Thus they have to stop at the first pre-signal. In order to obtain better elasticity in traffic flow, funnels with two speed signals were installed. Although two such funnels are already installed in Düsseldorf, nothing as yet can be said as to their precise degree of efficiency because they could not yet be put in service on account of construction work in progress nearby. Their performance can, however, be predicted according to general observations.

The special significance of the second speed signal lies in the fact that between the last vehicle of a group and the first vehicle of the following group there is a time gap in which no speed can be indicated because too great a speed difference would result in too short a time. In order to make the driver aware of how far he is outside the time stream, he is shown an amber flashing signal at this spot. In this manner he is warned that he may be forced to slow down at the next signal (Fig. 12).

It may be mentioned here what trifling effects influence correct behavior at pre-signals. Some years ago, it was observed that on a certain road all vehicles stopped when arriving at a certain pre-signal during the red period. A considerable part of all vehicles passed another pre-signal not far away during the red period. An investigation of this discrepancy led to the finding that at the first pre-signal there was a stopping line marked in front of it, while at the other signal there was none. After such a line had been applied there also, all vehicles stopped when the red light was shown at this signal.

In urban areas it is recommended to install the pre-signals before junctions, if possible, even if this would lead to an irregular displacement of the signal. The advantage lies in the improved traffic flow from the junction and the possibility for pedestrians to cross the street. Pre-signals must always be visible also from the overtaking lane (repetition of the signal on a whip pole), in order to avoid surprise reactions.

FUNNELS CONSISTING OF SPEED SIGNALS ONLY

On an important highway with dense traffic two funnels in both main directions were installed before a signalized intersection. These consisted of one speed signal

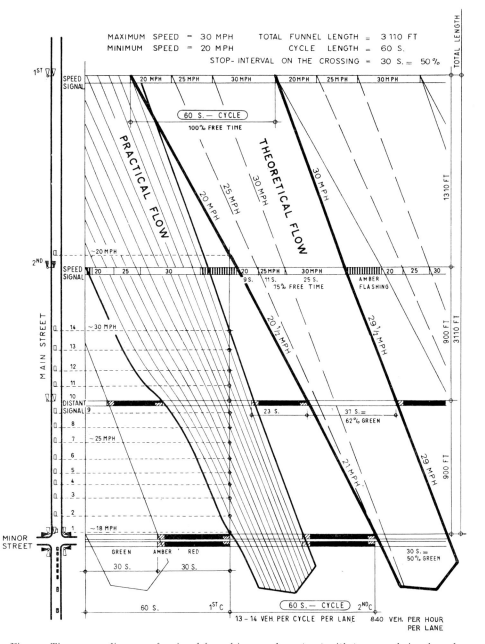

Fig. 12. Time-space diagram of a signal funnel in an urban street with two speed signals and one pre-signal.

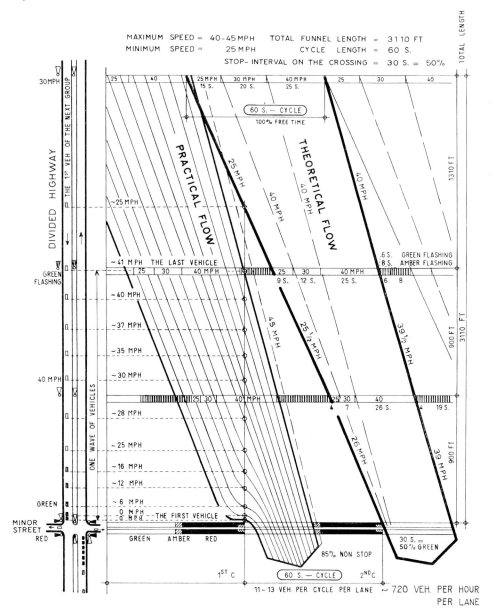

Fig. 13. Time–space diagram of a signal funnel on a highway with speed signals only.

and two pre-signals each over a total distance of about 2800 ft. Although the efficiency of traffic flow was very good, with free flow of 95 to 98%, it was observed that 2 to 5% of the vehicles had to stop at the pre-signals. It also happened that there were no stops for more than one hour. On rare occasions accidents were observed, caused by vehicles crashing into the rear of vehicles braking suddenly, an effect known from all signal installations. While accidents of this type usually are only of slight consequence in urban traffic due to lower speeds, they are considerably more dangerous on open highways. This experience has probably been the reason why repeated warnings have been voiced against the application of light signals on open highways.

For roads on which higher speeds are permitted traffic funnels consisting only of speed signals (Fig. 13) are therefore advisable. On long stretches these funnels lead to an adjustment of speeds, at the same time avoiding full stops, except before the intersection itself.

When the last vehicles of a group pass the speed signal shortly after the extinction of the maximum speed of the funnel (40 mph) they might be forced into the following period by sudden braking. To avoid this, the dark pause between 40 and 25 mph should start with a green light flashing briefly. It then changes to an amber flashing light as soon as the highest admissible speed of the road (45 mph) is surpassed. This method should give the greatest safety and also should be favorable to traffic flow. It must, however, be kept in mind to limit maximum speed on the road to 45 or 50 mph by signs at least 1000 ft before the funnel.

SHORT FUNNELS TO EQUALIZE DIFFERENT SPEEDS IN GREEN WAVES

When streets are irregularly intersected, the usual method of obtaining a straight time–space band for intersections displaced from the dividing point has been by lengthening the green period. The dividing point is the location in the time–space band where the traffic streams of opposite direction meet. A signal placed at this dividing point has simultaneously the greatest amount of green time for cross traffic. This means that main intersections must be placed at dividing points. Intersections with less cross traffic may be more distant from a dividing point if their cross traffic is smaller. Fig. 14 shows such a straight time–space band with constant speed, with intersection number 2 at a distance of 100 ft and intersection number 3 at a distance of 150 ft from the dividing point. All 4 intersections have an equal green time of 44 s for the main directions. At intersections number 1 and 4 there remain 24 s for cross traffic, at intersection number 2 already only 20 s, and at intersection number 3 not more than 18 s of green time. If the traffic density at intersections displaced with respect to the dividing point does not require longer green periods, then everything is in order. However, if the cross

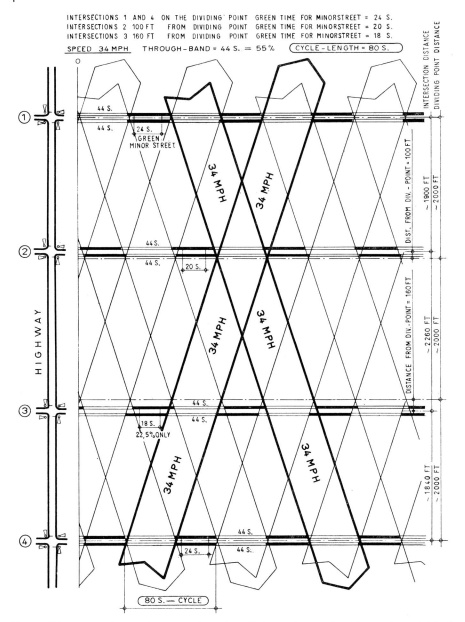

Fig. 14. Time–space diagram by unequal spacing of intersections and rectilinear through-band.

traffic is more dense, some traffic engineers suggest varying speeds. Speed changes of 2 mph in decelerating and 3 mph in accelerating from the normal speed of the through band, will generally be accepted by experienced drivers. But greater variations of speed will invariably lead to the forming of queues. In urban areas these queues are merely disagreeable and detrimental to traffic flow efficiency. In suburban areas with higher speeds the sudden disintegration of a group of vehicles endangers safety, because the damages resulting from crashes at speeds over 30 mph are usually considerable.

To avoid these shortcomings, short funnels were developed which by means of speed signals ostensibly indicate the irregular speeds suggested by the circumstances. Although they were only meant as friendly recommendations, which drivers need not necessarily obey, the improvement of traffic flow was unexpectedly great. The majority of drivers willingly adapted themselves even to speed variations of 8 to 10 mph.

The most important success of these short funnels between intersections lies in the fact that when the cycle changes from green to red, that is, at the amber light, there are in general no vehicles at all before the intersection. Thus the possibility of unexpected brakings is eliminated to a large extent. This dead zone occurs through a slight increase in formation density due to the fact that the first drivers are slowed down and the last drivers are speeded up slightly by the speed signal. This increase in formation density need not be reflected in a substantially greater proximity of the vehicles to one another, when the traffic is not saturated. For instance if the first three vehicles of a group drive at 30 mph they will be caught up with, not very far from the next intersection, by the following three vehicles using the overtaking lane, since the signal has indicated a speed of 35 mph to these last vehicles. In this manner enough space is gained for the rest of the group driving at 40 mph.

This method has been very much appreciated by drivers and also has proved successful with regard to safety. The gain in green time for intersecting traffic has in many cases enabled the planning of a green wave which otherwise would have been difficult to introduce.

Fig. 15 shows how under the same circumstances as in Fig. 14 the time–space band between intersections numbers 1 and 2 as well as numbers 3 and 4 is slowed down to 30 mph and the one between intersections numbers 2 and 3 is speeded up to 38 mph. Consequently the green time for intersecting traffic could be increased from 20 to 24 s at intersection number 2 and even from 18 to 24 s at intersection number 3. This may be decisive for traffic flow efficiency during peak hours.

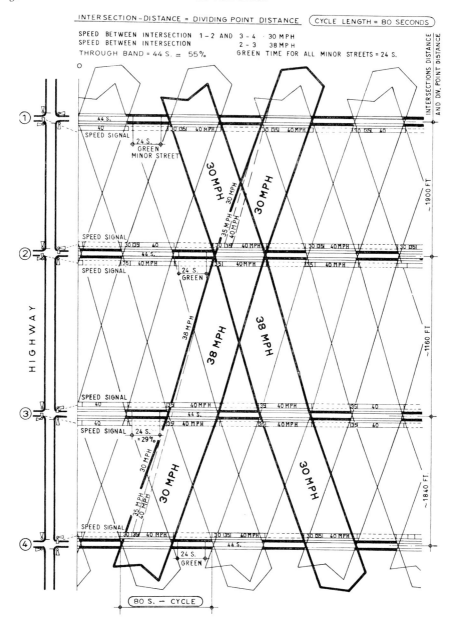

Fig. 15. Time–space diagram by unequal spacing of intersections and nonrectilinear through-band with speed signals between intersections.

INTERMEDIATE FUNNELS FOR TOO GREAT DISTANCES BETWEEN INTERSECTIONS

It is an acknowledged experience that at cycles of more than 60 s the next signal brings about the proper bunching of a green wave when showing the opposite signal light phase. If there is no intermediate intersection between the points with corresponding phases, a marked dispersion of the groups of vehicles will be observed. In such situations the author originally used to bridge the gap by ordinary pre-signals. When the distance between signals is great, a more efficient solution is to place speed signals immediately after the intersection. In extreme cases additional speed signals are placed halfway between intersections. On speed-ways without adjacent habitation these intermediate funnels have an important safety effect, just because the vehicles are squeezed away from the critical signal phase. Even when the time–space band is straight, intermediate funnels are always advisable when distances between signals are great.

THE EXTREMELY LONG FUNNEL

The longer the distance along which a funnel is planned, the less the variations in speed may be, and the softer and less forced the driving attitudes will be. Fig. 16 shows an extremely long funnel in Düsseldorf whose pleasant driving conditions have found wide appreciation. In its last third (500 m) before the intersection the first original funnel was installed in 1954.

Before the main intersection (Auf'm Hennekamp) at a distance of 650 m there is a cross walk of medium traffic density for pedestrians at Moorenstrasse. Another 540 m before it, there is another very narrow crosswalk at Christoph-strasse. The latter had become very dangerous before signals were installed. In the nine-month period before their installation four fatal pedestrian accidents had occurred. After the installation of the long funnel, which has been in use for two years, only one fatal accident has occurred, and this one far outside the cross-walk area.

Steadily increasing cross traffic times in consecutive intersections demand the installation of an extremely long funnel which itself is intersected twice, so to speak. Such a layout is not even rare, because in approaching an area of greater density of traffic, streets with increasing cross traffic are bound to become more numerous. Therefore, this example is deemed to be of principal importance.

The opposite southbound funnel, beginning 310 m before the Moorenstrasse, is also quite interesting. Here the time–space band is considerably widened. This fact can be utilized for inserting into the traffic flow a considerable number of vehicles. They proceed from west to south by a right turn at the main inter-section. The comparatively long cycle of 80 s is a consequence of speeds and

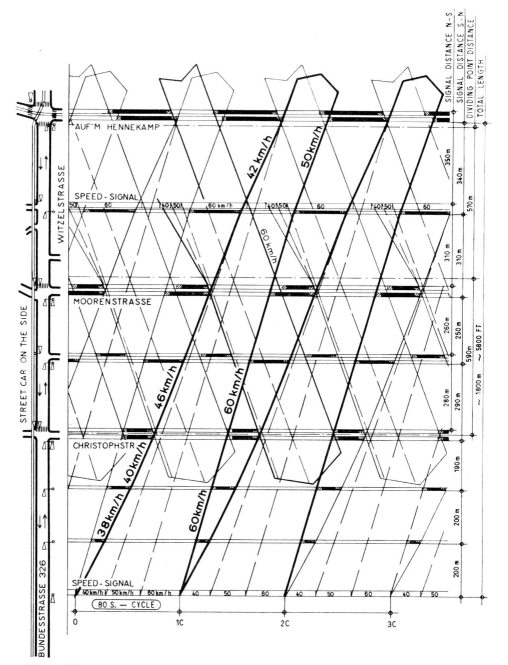

Fig. 16. Example of a long signal funnel with pedestrian crossings inside the funnel in Düsseldorf (Germany).

distances of dividing points. They are also a consequence of the complicated conditions at the main intersection where the street car moves from the center of the road to the side of the road. At the same time the traffic turning left from south to west toward an important Rhine bridge makes an additional left phase necessary.

THE COORDINATION OF INTERSECTIONS WITH DIFFERENT CYCLES BY MEANS OF TRAFFIC FUNNELS

For traffic engineers it has always been an almost impossible task to synchronize several coordinated systems with differing cycles. The development of the traffic funnel today affords in some cases smooth traffic flow in spite of intermediate intersections with differing cycles.

Fig. 17 shows the time–space diagram of a case in which the coordination of a 50 s cycle with a 64 s cycle has been developed. The only condition is the existence of a distance between intersections which is sufficient for the installation of a traffic funnel. On account of the many different speed lines the readability of the diagram was improved by separate charting of northbound and southbound traffic. This method, too, has been tried in several places with success. It improves considerably the possibility of coordination.

THE INDIVIDUAL SIGNAL FUNNEL WITH TRAFFIC-ACTUATED CONTROL

According to its principles, the signal funnel seems to be applicable only for fixed time operation. This, however, is not so, though traffic flow efficiency with fixed-cycle control surpasses considerably the results obtained by using a normal traffic-actuated control system.

If, for instance, the main traffic stream of a road is very dense, but intersecting traffic rather slight, the main traffic will normally be privileged, and intersecting traffic will be arrested by a stop sign. As long as the main traffic stream has intermittent gaps, this type of regulation will be sufficient. However, it becomes impossible at a certain traffic density, because there are no longer any adequate time gaps. In these cases such a gap must be artificially formed by a signal. Until now this usually was actuated by traffic from the intersecting road, thus interrupting the main stream for a short period. At urban speeds of 30 mph this method works rather reliably and safely. But when there is no speed limit on the main road, signals which unexpectedly interrupt the flow become very dangerous. Here the individual signal funnel may help. At a sufficient distance from junctions or cross streets with light traffic, speed signals for only two speeds, for instance 30 and 40 mph, are placed. If possible these are repeated halfway to the intersection. When there is no intersecting traffic these signals constantly indicate 40 mph.

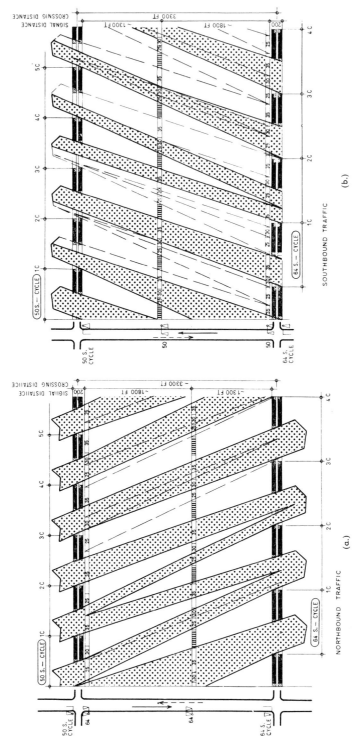

Fig. 17(a) and (b). Combination possibilities of intersections with different cycles by traffic funnels. (a). Time–space diagram for the traffic movement from 50 second cycle intersection to 64 second cycle intersection. (b.) Time–space diagram for the traffic movement from 64 second cycle intersection to 50 second cycle intersection.

When a vehicle from an intersecting street or a pedestrian actuates the controller, the pre-signals indicate the lower speed of 30 mph. As soon as all vehicles within the funnel have passed the intersection at 40 mph, the signal for the main direction changes to red. Those for the intersecting street change to green for a short period. Shortly before this the speed signals have again changed to 40 mph and maintain this speed indication until intersecting traffic again actuates the controller.

APPLICATIONS OF THE SIGNAL FUNNEL

In urban areas a signal funnel is always recommended when a coordinated signal system begins with an important intersection. Without such a funnel the loss of starting time during peak hours causes considerable disturbances. The funnel keeps traffic flowing already at the first intersection. This is why the funnel is increasingly used also at intersections which are not of extreme traffic density. If in a suburban area a single intersection signal is installed at the crossing of two arterial roads, at least two funnels in the main direction are indicated for through traffic with asymmetrical traffic density. Four funnels are advisable, if the main road and the intersecting road have approximately equal traffic densities.

Intersections of highways or through-roads in small towns are another field for the application of traffic funnels.

In areas near city limits it often happens that drivers have to pass from one wave system to another with a different cycle. A smooth changeover to the new system can be facilitated through the use of signal funnels.

On high-speed roads, *i.e.* over 30 mph, the signal funnel is indispensable for the safe operation of a signal system because the average driver lacks experience about braking distances at high speed and since usually no sudden braking is necessary on stretches where high speeds are possible.

A few words about the so-called dial signals or traffic dials which were intended to influence traffic flow by dial indication of time. Besides other unfavorable peculiarities which caused ultimate abolition of this type of signal, a mere indication of time does not suffice. The average driver does not know which speed he has to maintain when a certain amount of time is indicated on the dial. Measurements show that dial signals caused an increase in traffic flow efficiency of about 6% only, while signal funnels are apt to bring about almost completely uninterrupted flow of traffic. The speed signal indicates the right speed at the right place, thereby virtually eliminating all mental effort on the part of the driver.

THE ECONOMIC SIDE OF TRAFFIC FUNNELS

Expenditure for cables and signals to install traffic funnels may appear somewhat high at first, when it is only intended to improve traffic flow. On high-speed roads it simply becomes a necessity for safety reasons, which does not permit one to consider costs. Nevertheless it can be shown that a signal funnel pays for itself. This advantages lies in avoiding the many stops. According to findings in Europe it takes the same amount of gas to accelerate a stationary passenger car to 30 mph as to drive it 0.6 miles at 30 mph. Assuming a medium traffic load of 800 to 900 vehicles per hour, this amounts to 250 vehicle-miles per hour, considering that about half of the vehicles would have to stop in the absence of a funnel. On a normal day this would add up to a fuel consumption of 4000 car-miles.

Appendix

After my lecture in Detroit I was asked to what extent traffic regulation by a signal funnel has found use in Germany. As far as I know, there are at present about 40 full signal funnels, but a great many more intermediate funnels between coordinated signals. While Düsseldorf has only 22 of these funnels so far, Cologne already has 160. In Cologne the principle has been adopted to install a speed signal after each intersection which is coordinated with the next one. Düsseldorf and Essen intend to follow this example. This procedure has another advantage as it indicated to every driver, especially those from out of town, which signals are coordinated to each other and which speed to maintain. Speed indicators in the form of a fixed metal speed sign such as are also in use in the United States have the disadvantage that they indicate the average speed only when the time–space band is straight. In order to have a coordinated signal funnel, it is not possible to avoid making changes in the indicated speed in the time–space band. In this case the speed signal always indicates the correct speed. Speed signals are especially favorable when the wave speed is increased in the evening. In Cologne speed signals are estimated for another reason. By varying speeds within the time–space band free time for cross traffic can be increased without time losses for the main traffic stream. In Essen and Cologne it is planned to put full signal funnels ahead of every coordinated signal system. These advantages and the extensive application of the funnel principle in our region will probably make for increased adoption of this system elsewhere.

The Distribution of Traffic on a Road System

J. G. WARDROP, B.A., F.S.S.

*Road Research Laboratory, Department of Scientific and Industrial Research
Harmondsworth, West Drayton, Middlesex, Great Britain*

ABSTRACT

The problems of the distribution of traffic on a road system and the relation between the demand for travel and the cost of transport are considered. A theoretical model is proposed for passenger movements in which the average number of trips per household per unit time to each destination i is proportional to the function $A_i z_i^{-1} e^{-\lambda z_i}$, where A_i is a measure of the attraction of i, and z_i is the total cost of a return trip to destination i. In determining costs for this purpose both direct costs (*e.g.* fares, fuel) and indirect costs (*e.g.* time consumed, discomfort and accident risk) should be included as far as possible. The average expenditure E_i on trips to i per unit time is then proportional to $A_i e^{-\lambda z_i}$, and if E is the total expenditure per unit time,

$$E_i = \frac{A_i e^{-\lambda z_i} E}{\sum_i A_i e^{-\lambda z_i}}$$

In general E is assumed to be a known function of the average cost of a trip $\bar{z}$, given by

$$\bar{z} = \frac{\sum_i A_i e^{-\lambda z_i}}{\sum_i A_i z_i^{-1} e^{-\lambda z_i}}$$

In particular, it may be assumed as an approximation that E is independent of $\bar{z}$. In the case of journeys to work in Greater London in 1953–4 the value of λ is approximately given by $\lambda^{-1} = 3.75$ shillings. The total expenditure on travel in Greater London in the same year was 44 shilling per household per week. If the value of λ defined above applied to all journeys, and the distribution of 'attractors' were uniform, the number of trips per week per household would be $44/3.75 = 11.7$. The actual number per household per week (excluding journeys by foot) was recorded as 13 by a sample survey in 1954.

The theoretical model is applied to a few simple special cases in addition to that of a uniform distribution of attractors, including the case of two towns. The effects on traffic of building a motorway with access points at each end (along which the cost of travel per unit distance is reduced) are discussed for two cases, (i) where the cost via existing roads is proportional to the 'airline' or direct distance, (ii) where the cost is proportional to the distance along a rectangular grid.

References p. 77/78

INTRODUCTION

The way in which road traffic distributes itself on a network of roads is a matter of great importance to traffic engineers. It affects the volume of traffic which will use a new or improved facility, the benefits which are derived from improvements and hence their economic justification, and the traffic volumes which are expected in the future. A great deal of research has been done on the problems of traffic assignment and many theoretical models have been suggested for predicting traffic flows between cities or other traffic generators (see for example SCHIMDT AND CAMPBELL[1], BEVIS[2], the *Melbourne and Metropolitan Planning Scheme*[3] and FEUCHTINGER AND SCHLUMS[4]). Some of these models assume that the total number of trips originating in each zone and the total number terminating in each zone are known; in this case the only problem is to allocate inter-zone trips to satisfy the totals and to take some account of distance between zones or other 'deterrent' factors. Other models are of the general form

$$N_{ij} = P_i P_j f(z_{ij})$$

where N_{ij} is the number of trips between zones i and j, P_i, P_j are the populations of zones i, j respectively and z_{ij} is a quantity related in some way to the cost of travel between the zones–the distance between them, the journey time between them or some other quantity. The function $f(z)$ is commonly of the form z^{-p} where p may have any value between 0 and 3, depending on the type of traffic and district considered.

The present paper considers a theoretical model which differs in some respects from any of the models so far mentioned and which is based on the work of TANNER[5]. The model is applied to some simple idealised traffic situations, to illustrate the possibilities of such an approach. It is felt that some useful general information on the pros and cons of alternative traffic plans could be obtained by the examination of such simple examples.

THE TRAFFIC MODEL

General

The model which is to be described relates to transport in very general terms. It is assumed that there is a demand for the movement of people and goods which is not fixed but depends on the cost of travel (see BECKMANN, McGUIRE AND WINSTEN[6]). In principle, 'transport' is taken to mean movement by any available means. In most cases of interest to highway engineers, however, only journeys by road or rail need be considered. It is assumed that any particular journey is made by the cheapest form of transport available. In determining transport costs for this purpose the direct costs (*e.g.*, the fare in the case of public transport, the running cost of the car in the case of private transport) and all relevant indirect costs (*e.g.*, the cost of time consumed, discomfort, and accident risk) should be included as far as possible.

For simplicity, the model will be limited to passenger journeys. Every journey is assumed to consist of a return trip from the home to some destination and back. Thus, all the (return) journeys undertaken by any individual are assumed to radiate from his home. Although the description is in terms of passenger movement, the same kind of model can be applied to goods movement as well. In the case of goods, the origin is the place of production or storage, the destination is the place of delivery and the second half of the journey is the return of the empty vehicle.

This very simple model omits round-trip journeys (which are particularly important in the case of goods delivery). The inclusion of round-trip journeys would involve serious difficulties since the choice of route for maximum economy is a complex process and it is therefore simpler to exclude them.

Traffic generators

It is assumed that there is a known distribution of population over the area being studied, the population density being $P(x, y)$ where (x, y) is a typical point in a Cartesian co-ordinate system. It is also assumed that the characteristics of the population, apart from its density, are uniform throughout the region (although this restriction could be relaxed if necessary). There is also a distribution of 'attractors' of traffic, with density $A(x, y)$. These attractors may be jobs, shops, places of entertainment, schools, residences, etc. Their density is measured in terms of the degree of attractiveness. The total attraction of an area is given by the integral

$$\iint\limits_{Area} A \, \mathrm{d}x \, \mathrm{d}y$$

If two small areas have total attractions A_1 and A_2, and if X is a point such that the cost of travel from X to each area is the same, then the traffic from X to the two areas is in the ratio $A_1 : A_2$.

Travel cost

It is assumed that the minimum cost of travel between any two points (x, y) and (X, Y) can be determined. We may call this the cost function

$$z = z(x, y, X, Y)$$

The simplest example of a cost function is that in which the cost is proportional to the 'airline' or direct distance between the points, so that

$$z = u \{(X - x)^2 + (Y - y)^2\}^{\frac{1}{2}}$$

where u is a constant (the cost per unit distance).

In real situations the cost function may be difficult to measure and is likely to be very complex. A major component of cost, particularly in congested urban areas, is the value of time spent travelling. This is proportional to the reciprocal of the speed, which can vary considerably within a given area. In $8\frac{1}{2}$ square miles of Central London the average speed during the working day varies from street to street over the approximate range of 5 mph to 20 mph. Assuming 3d. per mile plus time costs at 2s. 6d. per hour (as suggested by TANNER[5]) for travel by private car the total cost of travel in Central London varies in an irregular manner between 9d. and $4\frac{1}{2}$d. per mile. However, in a simple theoretical model much of this variation might be smoothed out. For instance, in a model of a city we might suppose that the speed had a certain fixed value throughout the central area and another higher value in the rest of the city. It would also simplify the model if the cost per unit distance is assumed to apply to the airline distance, not the distance actually travelled. More complicated assumptions can be made when the need arises.

The demand function

In a particular instance, we may be concerned with a specific class of transport (e.g., travel by private car, all travel by road) or possibly with all travel. Whatever the class of transport, it may be assumed that there is a demand function which is related to the average cost of travel. This function is discussed by BECKMANN, McGUIRE AND WINSTEN[6].

In order to specify a demand function for a commodity it is necessary to choose some units in which the commodity is measured. It might be thought that the logical unit for passenger transport would be the passenger-mile. On the other hand, as GOODEVE[7] has pointed out, passenger-mileage is not an end itself; he described the objectives of transport as 'meetings'. Bearing this in mind, the 'trip' seems to be the appropriate unit, despite the wide variation in the cost of alternative trips. It is a natural extension of this idea to say that the number of trips is a function of the average cost per trip, or

$$N = F(\bar{z})$$

where N is the number of trips per head from a given origin per unit time and $\bar{z}$ is the average cost per trip.

References p. 77/78

Deterrence functions

It will also be assumed, following TANNER[5], that the relative attractiveness of two destinations, with intrinsic attractions A_1 and A_2 and with costs z_1 and z_2 from a given origin O, are $A_1 f(z_1)$ and $A_2 f(z_2)$, where $f(z)$ may be called a 'deterrence function'. This means that if from a given origin O we have n alternative destinations with attractions A_i and costs z_i $(i = 1, 2, \ldots . n)$, the number of trips per unit time from O to destination i, N_i say, is proportional to $A_i f(z_i)$, i.e.

$$N_i = k A_i f(z_i)$$

where k is a constant to be determined.

Summing over i we have

$$N = k \sum_i A_i f(z_i) \qquad \text{or} \qquad k = \frac{N}{\sum_i A_i f(z_i)}$$

Average cost per trip

The average cost per trip is given by

$$\bar{z} = \frac{\sum_i N_i z_i}{\sum_i N_i} = \frac{\sum_i A_i f(z_i) z_i}{\sum_i A_i f(z_i)}$$

This depends only on the characteristics of the destinations available and not on the total number of trips. The total number of trips can therefore be determined from the relation $N = F(\bar{z})$.

Possible demand functions

A typical demand function for trips might be of the form

$$N = B(\bar{z})^{-p}$$

This has an elasticity of p in the standard terminology of economics.

TANNER[5] has studied the total expenditure on travel in various areas of Great Britain in 1953–54 as given by a survey of household expenditure. He has shown that in all areas, both rural and urban, the expenditure on travel per household, including an allowance for travelling time at an average rate of 2s. 6d. per hour, and adjusting where necessary for differences in income distribution, was substantially constant. The average weekly expenditure on travel was generally about £1.8 per household (or about 15 per cent of the total expenditure) but was a little higher in Greater London, about £ 2.2 per household. TANNER therefore suggests that it might be assumed that the total expenditure on transport (time and money) depends only on income and not on the cost of travel.

This is a special case of the kind of demand function mentioned above, in which the elasticity is unity. We then have

$$N = \frac{E}{\bar{z}}$$

where the constant E is in fact the total expenditure on transport per unit time.

A possible deterrence function

TANNER has also studied the relative frequencies of journeys to work over different distances, based on the 1951 census of population for England and Wales. He has fitted a function of the form

$$n_{ij} = a_i b_j \, e^{-\lambda r_{ij}} \, (r_{ij})^{-n}$$

to the number n_{ij} of persons living in zone i and working in zone j, where r_{ij} is the distance between the two zones. He has shown that there are theoretical advantages in a deterrence function of the form

$$e^{-\lambda r} \, r^{-n}$$

if $n < 2$. In the case of a uniform distribution of attractors, the travel from O to points at distances in the range $(r, r + dr)$ is proportional to $e^{-\lambda r} r^{1-n} \, dr$. If r is small this is approximately equal to $r^{1-n} \, dr$. The total number of journeys between O and r is then approximately

$$\int_0^2 r^{1-n} dr \; .$$

This is finite if $n < 2$. On the other hand, as r approaches infinity the number of journeys between r and $r + dr$ tends to zero as it should, provided that $\lambda > 0$. It can easily be seen that no value of n could give satisfactory results both as r tends to zero and as it tends to infinity for a deterrent function of the simpler form r^{-n}.

TANNER found from his analysis that for journeys to work from several sources a deterrent function of the form $e^{-\lambda r} r^{-n}$ would fit the data remarkably well. He also found different values for λ and n in different areas, but they usually approximately satisfied the relation $\lambda = 0.3 - 0.1 \, n$ where λ is measured in $(miles)^{-1}$. Fig. 1 illustrates four functions of this form, for which $n = 0, 1, 2$ and 3, and λ is given by the above expression. It will be seen that over the range of 2 to 25 miles, which covered most of the journeys studied, the four curves are fairly similar. TANNER suggested that the particularly convenient value $n = 1$ might be used, the value of λ being found by fitting to the available data.

References p. 77/78

Following this suggestion, but replacing the distance r by the cost of a *return* trip z, the following deterrence function is proposed

$$f(z) = e^{-\lambda r} z^{-1}$$

Here of course λ must have the dimensions of (cost)$^{-1}$. In terms of the original units, λ^{-1} for the journeys to work was approximately 5 miles (corresponding to 10 miles for the return trip). The average cost of travel (including time costs) deduced by TANNER in his investigation of expenditure came to 4.5 d. per mile for large towns (by public or private transport) and 4.0 d. per mile by private

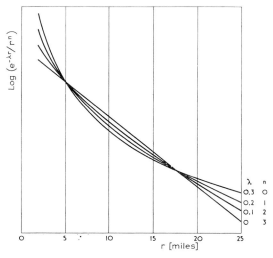

Fig. 1. Comparison between curves of log (e^{-r}/r^n) for $n = 0, 1, 2$ and 3. Values of λ chosen to give about the same slopes over the range studied. Origin of vertical scale different for each curve, so as to make curves approximately coincide.

transport, 4.5 d. per mile by public transport in rural areas. Applying the factor of 4.5 d. per mile we have

$$\lambda^{-1} = 45 \text{ d.} = 3.75 \text{ shillings.}$$

EXAMPLES OF APPLICATIONS

The infinite uniform distribution of attractors

The case of an infinite uniform distribution of attractors has already been mentioned. With the present deterrent function $e^{-\lambda r} z^{-1}$ we have to bring in the cost per unit distance, u, say. This will refer to the cost of a return journey from the origin to a given distance. If u is constant, the cost of a journey over a distance r and back is ur. In terms of r, the deterrence function is then $e^{-\lambda ur} r^{-1}$. (The multiplying constant can be dropped.)

References p. 77/78

The number of trips over the range of distance $(r, r + dr)$ is proportional to $e^{-\lambda u r}$ and the expenditure on these trips is proportional to $r\,e^{-\lambda u r}$. The total expenditure is E and hence the amount spent on journeys between r and $r + dr$ is

$$\frac{E\,r e^{-\lambda u r}dr}{\displaystyle\int_0^\infty r e^{-\lambda u r}dr} = E\,\lambda^2 u^2 r\,e^{-\lambda u r}dr$$

The average distance travelled is given by

$$\frac{\displaystyle\int_0^\infty r e^{-\lambda u r}dr}{\displaystyle\int_0^\infty e^{-\lambda u r}dr} = \frac{\mathrm{I}}{\lambda u}$$

and the average cost of a trip is evidently I/λ. If values of λ^{-1} and u appropriate to journeys to work in large urban areas (*i.e.* $\lambda^{-1} = 3.75$ sh. and $u = 9$ d. per mile) are applicable, the average outward distance travelled is 5 miles and the average cost per trip is 3.75 sh. The average weekly expenditure per houshold on transport in Greater London in 1953/54 was £2.2 (44 sh.) so that the average number of trips was $44/3.75 = 11.7$ per week. It is interesting to compare this with the average number recorded in a sample survey[8] of travel in London in 1954 which came to about 13 trips per household per week.

Two towns

Let us consider an entirely different situation, where there are two towns of given sizes a certain distance apart. Let the following symbols be defined.

	Town 1	town 2
Number of households	H_1	H_2
Attraction	A_1	A_2
Trips per week per household		
Internal	N_{11}	N_{22}
To other town	N_{12}	N_{21}
Total	N_1	N_2
Average length of internal journeys	a_1	a_2

Distance between towns	b
Cost/unit distance (throughout)	u
Total expenditure on transport per household per week	E
Deterrence function	$e^{-\lambda z}z^{-1}$
	(where z is cost)

Then
$$N_{11} = k_1 A_1\,e^{-\lambda u a_1}\,a_1^{-1} \qquad N_{22} = k_2 A_2\,e^{-\lambda u a_1}\,a_2^{-1}$$
$$N_{12} = k_1 A_2\,e^{-\lambda u b}\,b^{-1} \qquad\quad N_{21} = k_2 A_1\,e^{-\lambda u b}\,b^{-1}$$

References p. 77/78

where k_1, k_2 are constants to be determined. For origins in Town 1, multiplying the numbers of trips by the appropriate costs and adding gives

$$E = N_{11} (ua_1) + N_{12} (ub) = k_1 u (A_1 e^{-\lambda u a_1} + A_2 e^{-\lambda u b})$$

and therefore

$$k_1 = \frac{E}{u (A_1 e^{-\lambda u a_1} + A_2 e^{-\lambda u b})}$$

By a similar process k_2 can be found. It follows that

$$N_{11} = \frac{E A_1 e^{-\lambda u a_1} a^{-1}}{u (A_1 e^{-\lambda u a_1} + A_2 e^{-\lambda u b})} \qquad N_{12} = \frac{E A_2 e^{-\lambda u b} b^{-1}}{u (A_1 e^{-\lambda u a_1} + A_2 e^{-\lambda u b})}$$

$$N_{22} = \frac{E A_2 e^{-\lambda u a_2} a_2^{-1}}{u (A_1 e^{-\lambda u b} + A_2 e^{-\lambda u a_2})} \qquad N_{21} = \frac{E A_1 e^{-\lambda u b} b^{-1}}{u (A_1 e^{-\lambda u b} + A_2 e^{-\lambda u a_2})}$$

Since each trip is a return trip, the total flow from Town 1 to Town 2 is $H_1 N_{12} + H_2 N_{21}$ passengers per week and the flow in the opposite direction is the same.

The passenger mileages per week, bearing in mind that the distances quoted are half the totals covered in return journeys, are as follows

PASSENGER MILEAGE PER HOUSEHOLD PER WEEK

	Originating from	
	Town 1	Town 2
Internal	$2N_{11}a_1$	$2N_{22}a_2$
Between towns	$2N_{12}b$	$2N_{21}b$

As a specific example consider the following values

Town 1	Town 2
$H_1 = 5{,}000$ households	$H_2 = 20{,}000$ households
$a_1 = 1$ mile	$a_2 = 2$ miles
$A_1 = 1$	$A_2 = 2$

$$
\begin{aligned}
b &= 10 \text{ miles} \\
E &= 36 \text{ sh. per household per week} \\
\lambda^{-1} &= 3.75 \text{ sh.} \\
u &= 0.75 \text{ sh. per mile}
\end{aligned}
$$

Then

$E/u = 48$ miles (= half average passenger mileage per household per week)

$(\lambda u)^{-1} = 5$ miles

and

$$N_{11} = \frac{48 \times e^{-0.2}}{e^{-0.2} + 2e^{-2}} = 36.1 \text{ trips per household per week}$$

$$N_{12} = \frac{48 \times 2e^{-2}}{10\,(e^{-0.2} + 2e^{-2})} = 1.19 \text{ trips per household per week}$$

$$N_{22} = \frac{48 \times 2e^{-0.4}}{2\,(e^{-2} + 2e^{-0.4})} = 21.8 \text{ trips per household per week}$$

$$N_{21} = \frac{48 \times e^{-2}}{10\,(e^{-2} + 2e^{-0.4})} = 0.44 \text{ trips per household per week}$$

The passenger mileages are therefore as follows

PASSENGER MILEAGE PER HOUSEHOLD PER WEEK

	Originating from	
	Town 1	Town 2
Internal	72.2	87.2
Between towns	23.8	8.8
Total	96.0	96.0

The total flow from Town 1 to Town 2 is $5,000 \times 1.19 + 20,000 \times 0.44 = 14,750$ passenger movements per week, with the same number in the opposite direction. The internal traffic in each town is best described in terms of passenger mileage; it amounts to $5,000 \times 72.2 = 361,000$ passenger-miles per week in Town 1 and $20,000 \times 87.2 = 1,744,000$ passenger-miles per week in Town 2.

EFFECT OF VARYING COST PER UNIT DISTANCE

So far we have assumed that the cost per unit distance is constant. In practice, however, this cost can vary appreciably. We have seen that the cost of travel by private car in Central London during the day varies between 6d. and 1sh.3d. per mile. In many American cities there is a flat rate for travel by public transport, so that part of the cost per mile varies inversely with distance travelled. It follows that the theoretical model should be able to take account of variation in the cost per unit distance.

In many cases we may assume that the cost per unit distance is a function of position but not of the direction of travel. In such a situation it is no longer a simple matter to find the route from A to B which gives the minimun cost or the associated value of cost. Following BECKMANN, McGUIRE AND WINSTEN[6], we have to find 'isocost curves', which are the loci of points at a given minimum transport cost from a given origin. These can usually be constructed for small

References p. 77/78

increments of cost; each successive curve is derived from the previous one by finding all new points which can be reached from it for one increment of cost.

The system of curves orthogonal to the iso-cost curves for a given origin represent the paths along which the cost is minimal. If there are discontinuities in cost/unit distance these paths are refracted in a similar manner to rays of light at a discontinuity of refractive index. In general there is a very close analogy between the passage of traffic through an area of varying cost/unit distance and the passage of light through a medium of varying refractive index. Some interesting consequences of the 'refraction' of traffic have been discussed by KLAASEN AND NOU-HUYS[9].

A SINGLE SECTION OF MOTORWAY*

Let us assume the same model as before, with an infinite uniform distribution of attractors, and initially a constant cost per unit distance. To simplify the algebra we shall assume without loss of generality that the cost per unit distance is 1. Suppose that a motorway is built between A and B of length 2 units and with access only at the ends, and that the cost of travel along this motorway is μ per unit length ($\mu < 1$), so that the cost of travelling from A to B along the motorway is 2μ (see Fig. 2).

Iso-cost curves for one end of the motorway

Consider the iso-cost curves for journeys from one end of the motorway, A. Before the motorway is built the iso-cost curves are evidently concentric circles with centre A. After the motorway is built the iso-cost curve for cost z consists of part of the original circle, an arc of radius z centred at A, and a second arc of radius $z - 2\mu$ centred at B, if $z > 2\mu$ (see Fig. 2). If $z > 1 + \mu$ the two arcs meet on a hyperbola (the dashed curve in Fig. 2) which is the locus of points P such that $AP = BP + 2\mu$. For a point on this curve the cost of a journey is the same whether one travels via the motorway or not. For the area on the same side of the hyperbola as A it is cheaper to travel directly. For the area on the same side of the hyperbola as B, it is cheaper to travel via the motorway.

The effect of building the motorway is to lengthen the iso-cost curves for costs greater than 2μ. This is equivalent to increasing the amount of territory available at or below a given cost level.

The effect of the motorway on the lengths of the iso-cost curves

If $z < 2\mu$, the iso-cost curves are unaltered, their length being $2\pi z$. If $2\mu < z < 1 + \mu$, the length of the iso-cost curve is increased by $2\pi (z - 2\mu)$ by

* 'Motorway' is the British term for a limited-access road restricted to motor vehicles, which is equivalent to a freeway.

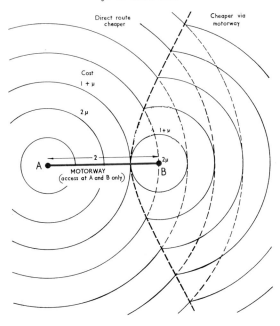

Assumptions:
    Cost/unit distance = 1 except on motorway
    Cost along motorway = $2\mu < 2$

Fig. 2. Iso-cost curves for one end of a motorway.

the presence of the motorway. For the effect on iso-cost curves for $z > 1 + \mu$ we must refer to Fig. 3. This shows the arc of the original iso-cost curve, PRQ, which is replaced by the new arc PSQ. With the angles θ and ϕ as shown in Fig. 3, the length of the new arc is $2 (z - 2\mu) \phi$. Hence the increase in the length of the iso-cost curves is $2 (z - 2\mu) \phi - 2 z \theta$. The angles θ and ϕ are given by

$$\cos \theta = \frac{z^2 + 4 - (z - 2\mu)^2}{4z} \qquad \cos \phi = - \frac{(z - 2\mu)^2 + 4 - z^2}{4 (z - 2\mu)}$$

or

$$\cos \theta = \mu + \frac{1 - \mu^2}{z} \qquad \cos \phi = \mu - \frac{1 - \mu^2}{z - 2\mu}$$

From these expressions it is a simple matter to calculate the increase in the length of the iso-cost curves for any value of μ. A change in the length of the motorway or in the value of cost/unit distance merely introduces a scale factor into the calculation.

References p. 77/78

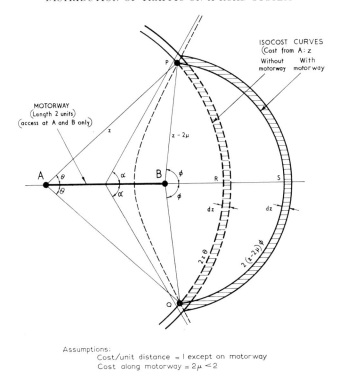

Fig. 3. Lengths of iso-cost curves for one end of a motorway.

The two shaded areas shown in Fig. 3 are as follows: (i) the area between the two dashed arcs which was previously in the cost range $(z, z + dz)$ and which is removed from that range; (ii) the area between the two solid arcs which comes into the cost range $(z, z + dz)$. Clearly the total increase in area in this range is $2 (z - 2\mu) \phi\, dz - 2z\theta\, dz$. This analysis draws attention to the effect of the motorway (or any improvement to transport facilities which reduces the cost of travel) in increasing the area within a given cost range.

Iso-cost curves for a general point

So far we have considered only journeys originating from A. Consider now a general point in the plane, X, where $XA = r$ and $XB = s$ (see Fig. 4). Suppose also that $r < s$. Let H_1 be the hyperbola corresponding to that in Fig. 2, but for journeys originating at B; it is defined by the equation $r = s - 2\mu$. Then if X lies on the same side of H as B (although nearer to A than to B) it is cheaper to travel from X to B directly than via the motorway. It follows that in this case no point in the plane can be reached more cheaply via the motorway. However, if X lies on the same side of H_1 as A (as in the case illustrated in Fig. 4), it is cheaper to travel to B via the motorway, the cost being $r + 2\mu$, whereas the direct cost is s, and in the case considered $r + 2\mu < s$.

We now have a situation very similar to the original one. Starting from X one can reach B more cheaply than by the direct route. The proportional saving in cost is

$$\frac{r + 2\mu}{s}$$

(corresponding to μ in the original situation) and the length XB is s (corresponding to 2). If we have tabulated changes in the lengths of iso-cost curves for a range

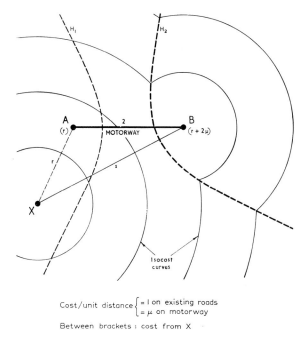

Fig. 4. Iso-cost curves from a general point X.

of values of μ, by making the appropriate change of scale we can find the effect on isocost curves for any point X. The iso-cost curves for X are drawn in Fig. 4.

When the general point X was introduced it was assumed that $r < s$. If $r > s$, by symmetry, the same results apply with A and B interchanged. If $r = s$ no use would be made of the motorway.

Clearly this analysis can be extended to cover the effect of the motorway on

References p. 77/78

the amount and character of the travel from X, using the deterrence function considered earlier. This is a straightforward computation which could be performed for a grid of points in the plane and for a range of values of μ and λ, using an electronic computer. Analysis of this kind is somewhat simpler if we use a model based on a rectangular grid of streets, and this will be considered in the next section.

THE RECTANGULAR GRID

In the models considered above the cost of travel on the existing transport system was assumed to be independent of the direction of travel. A model which

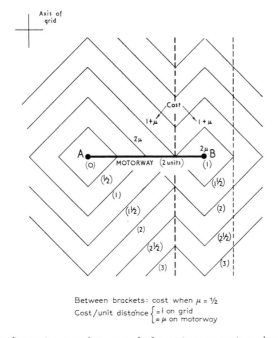

Between brackets: cost when $\mu = \frac{1}{2}$

Cost /unit distance $\begin{cases} = 1 \text{ on grid} \\ = \mu \text{ on motorway} \end{cases}$

Fig. 5. Iso-cost curves for one end of a motorway, rectangular grid.

is appropriate for many large urban areas is an idealisation of a rectangular grid of streets. If we have such a grid, with axes parallel to Ox and Oy and uniform speeds, the cost of travelling between two points on the grid (x, y) and (X, Y) is $u(|X - x| + |Y - y|)$, where u is the cost per unit distance travelled. If the mesh of the grid is small compared with the distances travelled we may neglect the condition that the points must lie on the grid and assume that the above cost function refers to any pair of points in the plane.

References p. 77/78

Iso-cost curves for a motorway

Consider the same simple case of a motorway AB, of length 2 units, parallel to one of the axes of the grid, and with access at A and B only (see Fig. 5). Assume that the cost per unit distance is 1 on the grid of existing streets and μ along the motorway. The iso-cost curves for A without the motorway are concentric diamond-shaped quadrilaterals. With the motorway the iso-cost curves are as shown in Fig. 5. The boundary of the area served by the motorway (the heavy dashed line in the diagram) is perpendicular to AB and at a distance $1 + \mu$ from A.

Changes in the lengths of the iso-cost curves

Inspection of Fig. 5 shows that the lengths of the iso-cost curves with and without the motorway are as follows

	Without motorway 4 $\sqrt{2}$ ×	With motorway 4 $\sqrt{2}$ ×
$0 < z < 2\mu$	z	z
$2\mu < z < 1 + \mu$	z	$2z - 2\mu$
$1 + \mu < z$	z	$z + 1 - \mu$

These lengths are plotted in Fig. 6; this graph also shows the subdivision of the iso-cost curve with the motorway into the length reached via the motorway and the length reached directly.

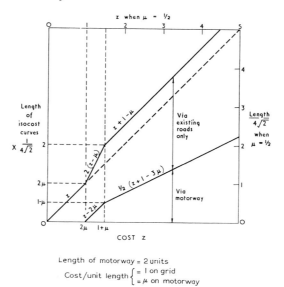

Fig. 6. Lengths of iso-cost curves from one end of a motorway, rectangular grid.

References p. 77/78

The effect of the motorway on the amount of travel from A

Let us assume the same deterrence function as before, $e^{-\lambda z} z^{-1}$, a constant expenditure on travel per household per week of E, and a uniform distribution of attractors. Before the motorway is built the expenditure on travel in the range of cost $(z, z + dz)$ is proportional to

$$e^{-\lambda z} z^{-1} z^2 dz = e^{-\lambda z} z dz$$

Since the total expenditure must equal E, the expenditure in the range $(z, z + dz)$ is

$$\frac{Ee^{-\lambda z} z dz}{\displaystyle\int_0^\infty e^{-\lambda z} z dz} = E\lambda^2\, e^{-\lambda z} z dz$$

After the motorway is built, the expenditure in this cost range is as follows (k being a constant to be determined):

Values of z	Expenditure
$0 < z < 2\mu$	$ke^{-\lambda z} z dz$
$2\mu < z < 1 + \mu$	$ke^{-\lambda z}\, 2\,(z - \mu)\, dz$
$1 + \mu < z$	$ke^{-\lambda z}\,(z + 1 - \mu)\, dz$

The total expenditure is

$$E = k\left\{ \int_0^{2\mu} e^{-\lambda z} z\, dz + \int_{2\mu}^{1+\mu} e^{-\lambda z}\, 2\,(z - \mu)\, dz + \int_{1+\mu}^\infty e^{-\lambda z}\,(z + 1 - \mu)\, dz \right\} =$$

$$= \frac{k}{\lambda^2} \{1 + e^{-2\lambda\mu} - e^{\lambda(1+\mu)}\}$$

Hence

$$k = \frac{E\lambda^2}{1 + e^{-2\lambda\mu} - e^{-\lambda(1+\mu)}}$$

A particular example

Values of expenditure in the particular case of $\mu = \frac{1}{2}$ and $\lambda = 1$ are shown in Fig. 7. With the actual values of λ and μ which have been used in earlier examples this corresponds to a motorway of length 10 miles. The unit of cost would be 3.75 sh. Fig. 7 shows that the effect of the motorway on travel from A is to reduce the expenditure on the cheaper (and shorter) journeys and increase the expenditure

References p. 77/78

on some of the more costly (and longer) journeys. It also shows that expenditure on travel via the motorway reaches a peak for journeys costing 1.5 units.

It is useful to calculate the number of trips via the motorway. We have the following values:

	Expenditure on trips via motorway	Number of trips via motorway
$2\mu < z < 1 + \mu$	$ke^{-\lambda z}(z - 2\mu)\,dz$	$ke^{-\lambda z}\left(1 - \dfrac{2\mu}{z}\right)dz$
$1 + \mu < z$	$ke^{-\lambda z}\tfrac{1}{2}(z - 3\mu + 1)\,dz$	$ke^{-\lambda z}\tfrac{1}{2}\left(1 - \dfrac{3\mu - 1}{z}\right)dz$

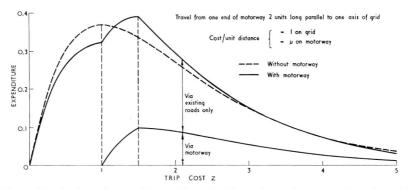

Fig. 7. Distribution of expenditure with and without the motorway, rectangular grid.

The total number of trips via the motorway is given by

$$k\left\{\int_{2\mu}^{1+\mu} e^{-\lambda z}\left(1 - \frac{2\mu}{z}\right)dz + \int_{1+\mu}^{\infty} e^{-\lambda z}\tfrac{1}{2}\left(1 - \frac{3\mu - 1}{z}\right)dz\right\}$$

$$= \frac{k}{\lambda}\left\{e^{-2\lambda\mu} - \tfrac{1}{2}e^{-\lambda(1+\mu)} - 2\lambda\mu I(2\lambda\mu) + \tfrac{1}{2}\lambda(1 + \mu)I(\lambda[1 + \mu])\right\}$$

where

$$I(y) = \int_{y}^{\infty} \frac{e^{-z}dz}{z} = -\mathrm{Ei}(-y)$$

Here $\mathrm{Ei}(y)$ is the exponential integral tabulated in the *Mathematical Tables*, Vol. I, of the British Association for the Advancement of Science[10].

An abbreviated table of $I(y)$ is given in Table I.

References p. 77/78

TABLE I.

$$\text{VALUES OF } I\ (y) = \int_{y}^{\infty} \frac{e^{-z}dz}{z}$$

y	I (y)	y	I (y)	y	I (y)
0.0	∞	1.0	0.219	2.0	0.050
0.1	1.823	1.1	0.186	2.5	0.025
0.2	1.223	1.2	0.158	3.0	0.013
0.3	0.906	1.3	0.136	3.5	0.007
0.4	0.702	1.4	0.116	4.0	0.004
0.5	0.560	1.5	0.100	4.5	0.002
0.6	0.454	1.6	0.086	5.0	0.001
0.7	0.376	1.7	0.075	5.5	0.001
0.8	0.313	1.8	0.065	6.0	0.000
0.9	0.263	1.9	0.056		

The total number of trips in a given cost range when the motorway is in use is as follows:

	Total number of trips
$0 < z < 2\mu$	$ke^{-\lambda z}\ dz$
$2\mu < z < 1 + \mu$	$ke^{-\lambda z}\ 2\left(1 - \dfrac{\mu}{z}\right) dz$
$1 + \mu < z$	$ke^{-\lambda z}\left(1 + \dfrac{1 - \mu}{z}\right) dz$

The grand total of trips is therefore

$$k \left\{ \int_{0}^{2\mu} e^{-\lambda z}dz + \int_{2\mu}^{1+\mu} e^{-\lambda z}\ 2\left(1 - \frac{\mu}{z}\right) dz + \int_{1+\mu}^{\infty} e^{-\lambda z}\left(1 + \frac{1 - \mu}{z}\right) dz \right\}$$

and this can be integrated to give

$$\frac{k}{\lambda}\{1 + e^{-2\lambda\mu} - e^{-\lambda(1+\mu)} - 2\lambda\mu I\ (2\lambda\mu) + \lambda\ (1 + \mu)\ I\ (\lambda[1 + \mu])\}\ .$$

Since $k = E\lambda^2/\{1 + e^{-2\lambda\mu} - e^{-\lambda(1+\mu)}\}$ the above expression can be simplified to

$$\lambda E + k\ \{- 2\mu I(2\lambda\mu) + (1 + \mu)I(\lambda\ [1 + \mu])\}$$

In our particular example, with $\mu = \frac{1}{2}, \lambda = 1$, we have

$$k = \frac{E}{1 + e^{-1} - e^{-1.5}} = 0.8735E$$

and the total number of trips after the motorway is built is

$$E + 0.8735E\ \{- I(1) + 1.5I(1.5)\} = 0.940\ E$$

References p. 77/78

Without the motorway the number of trips is λE (E in our case). Thus the presence of the motorway reduces the total number of trips from A by 6 per cent.

The number of trips on the motorway is

$$0.8735E \{e^{-1} - \tfrac{1}{2} e^{-1.5} - I(1) + \tfrac{1}{2}(1.5)I(1.5)\} = 0.098E$$

That is to say, of the traffic originating from A the motorway carries a flow equal to about 10 per cent of the original flow from A.

The expenditure on travel on the motorway itself is also $0.098E$, since each return trip costs 1 unit. The 'passenger-mileage' actually on the motorway (in the length units appropriate to this example) is $0.392E$. Summarising, we have the following relative figures for traffic originating from A:

	Without motorway	With motorway		
		On motorway	Not on motorway	Total
Trips	100	10	84	94
Expenditure	100	10	90	100
Passenger mileage	100	20	90	110

Similar calculations can be made for traffic originating from a grid of points in the plane, so that the whole effect of the motorway can be assessed.

One point which emerges from this analysis is that it is not easy to describe the process in terms of 'diverted' and 'generated' traffic. The total passenger mileage has increased by 10 per cent and this might reasonably be called generated traffic. On the other hand the effect of building the motorway (on the assumptions made in this paper) is to reduce the total number of trips and to alter the whole pattern of travel, so that we cannot describe the process in simple terms of trips with fixed origins and destinations being diverted to the motorway.

DISCUSSION

Effects of traffic congestion

So far traffic congestion has been ignored. In practice, particularly in large urban areas, it has a serious effect on the amount and distribution of traffic. The cost of travelling to a given destination is generally an increasing function of the flow of traffic on each part of the route, and this is a serious complication in the use of theoretical models. In principle it can be dealt with by starting with estimated degrees of congestion, calculating the resulting flow pattern, modifying the costs on the basis of the new flows, and repeating this process until it converges. The procedure would be rather involved, but would be feasible for simple models using an electronic computer.

Flow on a network of roads

Except for the example of two towns the previous discussion has been generally confined to idealised models in which travel can occur throughout the plane. The process can of course equally well be applied to travel on a network of roads. In this case, if we know the cost of travel along each link of the network it is necessary to calculate the minimum journey cost between any pair of points and the corresponding route. The problem of networks, including the effects of traffic congestion, has been discussed by PRAGER[11]. WHITING AND HILLIER[12] and DANTZIG[13] have given systematic methods of doing this which can easily be carried out on an electronic computer.

Prediction of future traffic

Most problems requiring traffic estimation and allocation are concerned with the future. This requires prediction of changes in traffic patterns produced by changes in population and land use generally, in the cost of transport, and in the distribution of income. The model described in this paper can be used to predict the effects of these changes. In particular, TANNER[5] has shown how the total expenditure on transport varies with income, and this relationship can be used to allow for income changes.

CONCLUSION

It has been shown that a plausible model for the relation between the amount of traffic and the cost of travel can be set up and applied to simple theoretical situations. By the use of electronic computers it would be possible to apply this model to explore the effects of new roads or other improvements in transport facilities on the quantity and quality of the transport of both passengers and goods in typical instances. It is to be hoped that some useful generalizations would emerge from such investigations.

ACKNOWLEDGEMENTS

The author is indebted to his colleagues at the Road Research Laboratory for their help in the preparation of this paper, and particularly to Mr. J. C. TANNER, whose work is the foundation on which the paper is built. The paper is published by permission of the Director of Road Research.

REFERENCES

1 R. A. SCHMIDT AND M. E. CAMPBELL, *Highway Traffic Estimation*, Eno Foundation for Highway Traffic Control, Saugatuck, Conn., 1956.
2 H. W. BEVIS, Forecasting zonal traffic volumes, *Traffic Quart., 10* (1956) 207–222.

3 *Melbourne Metropolitan Planning Scheme 1954*, Surveys and Analysis, 174–7, Melbourne and Metropolitan Board of Works.

4 M. E. FEUCHTINGER AND J. SCHLUMS, A traffic study concerning the arterial road system in the Munich Area, *Second Intern. Course in Traffic Eng.*, Burgenstock, 1954 (World Touring and Automobile Organization).

5 J. C. TANNER, Factors affecting the amount of travel, *Dept. Sci. Ind. Research, Road Research Tech. Paper No. 51*, H. M. Stationery Office, London, 1961. (In the press).

6 M. BECKMANN, C. B. McGUIRE AND C. B. WINSTEN, *Studies in the Economics of Transportation*, Cowles Foundation for Research in Economics, Yale Univ., New Haven, Conn., 1956.

7 C. GOODEVE, Man must measure, *J. Inst. Transp.*, 27 (1957) nr. 3.

8 *London Travel Survey 1954*, London Transport Executive, London, 1956.

9 L. H. KLAASEN AND J. H. NOUHUYS, *Traffic Problems and Town Planning*, Monograph No. IX of the 'Nederlands Verkeersinstituut', The Hague, 1954.

10 *Mathematical Tables*, Vol. I., British Association for the Advancement of Science, Cambridge Univ. Press, New York, 1956.

11 W. PRAGER, Problems of traffic and transportation, *Proc. Symposium on Operat. Research Business Ind.*, Midwest Research Inst., Kansas City, Mo., 1954, p. 105–113.

12 P. WHITING AND J. A. HILLIER, A method for finding the shortest route through a road network, *Operat. Research Quart.*, 11 (1960) nr. 1.

13 G. B. DANTZIG, On the shortest route through a network, *Management Sci.*, 6 (1960) 187–90.

The Relative Distribution of Households and Places of Work

A Discussion of the Paper by J. G. Wardrop

M. H. COHEN

Institute for the Study of Metals, University of Chicago, Chicago, Illinois

ABSTRACT

The problem of determining the dependence of the relative distribution of households and places of work on the cost of transportation between them is attacked by the methods of statistical mechanics, a branch of physics. The results agree with the model proposed by WARDROP to represent the statistical data. A refinement of the argument leads to a more flexible description of population distributions capable of representing, for example, a city, with the city proper, city limits, and suburbs emerging as natural features of the calculation. The principal goal, however, is not any specific result, but a demonstration of the utility of statistical mechanics for such problems.

In *The distribution of traffic on a road system*, a paper presented to this symposium, WARDROP[1] has put forward a simple model for the quantitative description of passenger movements. In this model, the average number of trips per household per unit time to each destination i is proportional to the function $A_i z_i^{-1} e^{-\lambda z_i}$, where A_i is a measure of the attraction of i, and z_i is the total cost of a return trip to destination i. The model is proposed by WARDROP on an empirical and heuristic basis as describing adequately the statistical information available, having desirable analytical features, and being readily applicable to various problems of distribution of traffic flow. The present note attempts to provide a deeper basis for WARDROP's model.

For definiteness and clarity, we limit our discussion to only one kind of return trip discussed by WARDROP, that from home to work within a well defined population area. Let us choose the working day as the unit of time and assume one worker per household. The average number of trips per household per unit time having total cost of transportation in the range between z and $z + dz$ is then identical to the fraction of all households in the population area for which the total cost of transportation to work is $f(z)\,dz$ in the range between z and $z + dz$. In the following, we attempt to derive this probability distribution $f(z)$ of households relative to cost of transportation to work z. We thus find ourselves addressing the problems of population distribution and economic behaviour underlying traffic flow rather than the problems of traffic flow itself.

Within a given population area the distribution function $f(z)$ varies with time. The locations of households and jobs change from time to time for each worker with an average interval between moves of, say, τ. The distribution function $f(z)$ therefore fluctuates during times comparable with τ. Large scale population shifts and economic changes cause slow variation of $f(z)$ over much longer times T. We are interested in $f(z)$ only after it is averaged over periods t long compared with τ and short compared with T. Provided the number of households is large and T is sufficiently larger than τ for suitable averaging periods t to exist, the average value of $f(z)$ does not differ appreciably from the value of $f(z)$ most frequently encountered in the interval t. It is therefore the most probable distribution which we calculate here by methods familiar in statistical mechanics, the branch of physics concerned with the thermal properties of matter[2]. Rather than refer the reader to a standard text, we choose to give the derivation explicitly here.

Again provided the number of households is large and T is sufficiently larger than τ for suitable averaging periods t to exist, the total population N of the area remains constant over t and so does the total cost of round-trip transportation per day Z for the entire population. We must therefore find the probability of a distribution $f(z)$ and maximize this probability subject to the restrictions that N and Z have their given, constant values.

To solve the statistical problem posed above, we first break the total range of values for z into regions of equal size Δz sufficiently small for $f(z)$ to be constant within such a region. The average value of z in the jth range is z_j, and the number of workers having z in the jth range is

$$n_j = f(z_j)\Delta z \tag{1}$$

The total number of workers N and the total cost of transportation Z may therefore be written

$$\sum_j n_j = N \tag{2}$$

$$\sum_j n_j z_j = Z \tag{3}$$

We now suppose that the number $\varrho(z)$ of possible locations for households per unit cost of transportation is a known function of z. The number of household sites in the jth range is therefore

$$\varrho(z_j)\Delta z = \varrho_j \Delta z \tag{4}$$

Finally, we suppose that there is a certain intrinsic preference for living close to work which we can characterize by the probability $g(z)$, with $g(z_j) = g_j$, that a worker chooses to locate his household on a given site.

We are now in a position to write down the relative probability or weight W of a given set of values $\{n_j\}$ of the n_j, $W(\{n_j\})$. Consider first a specific arrangement of workers on household sites. The probability that a given worker lives in the jth range is proportional to the number of sites in that range $\varrho_j \Delta z$ times the preference factor g_j provided that a worker is uninfluenced in his choice of a site by the presence of other workers. In that case, the probability of the specific arrangement of all the workers over their household sites is simply the product of the probability of finding each worker in the range he lies in

$$\prod_j (g_j \varrho_j \Delta z)^{n_j} \tag{5}$$

However, we are not interested in the probability that a specific group of n_j workers is in the jth range but only in the probability that any n_j workers are in the jth range. Hence, to get $W(\{n_j\})$ we must multiply Eq. (5) by the number of ways in which one can partition N objects into groups containing n_j objects each, which is

$$\frac{N!}{\prod_j n_j!} \tag{6}$$

Combining the factors (5) and (6) we obtain the desired result for W

$$W(\{n_j\}) = N! \prod_j \frac{(g_j \varrho_j \Delta z)^{n_j}}{n_j!} \tag{7}$$

We now proceed to find the set of n_j for which W, or equivalently $\log W$, is a maximum subject to the constancy of N and Z by the method of Lagrangian multipliers. Consider an infinitesimal variation $\{\delta n_j\}$ away from the set $\{n_j\}$ for which W is a maximum. The variation in W must vanish

$$\delta \log W = 0 \tag{8}$$

because it is a maximum, and so must those of N and Z

$$\delta N = \sum_j \delta n_j = 0 \tag{9}$$

$$\delta Z = \sum_j \delta n_j z_j = 0 \tag{10}$$

because they are constant. The variation of $\log W$ is readily evaluated by noting that

$$\frac{d \log x!}{dx} = \frac{\log (x+1)! - \log x!}{1} = \log x \tag{11}$$

for large x, or

$$\delta \log W = \sum_j [\log (g_j \varrho_j \varDelta z) - \log n_j] \, \delta n_j = 0 \tag{12}$$

Let us now multiply Eq. (9) by ν and Eq. (10) by $-\lambda$ and add the results to Eq. (12) to get

$$\sum_j [\log (g_j \varrho_j \varDelta z) + \nu - \lambda z_j - \log n_j] \, \delta n_j = 0 \tag{13}$$

Because of Eqs. (9) and (10) the variations of the δn_j are not independent, δn_1 and δn_2, say, being fixed by the remaining variations. If, however, we fix the values of the multipliers λ and ν so that the coefficients of δn_1 and δn_2 in Eq. (13) vanish, then the coefficients of the remaining independent and arbitrary variations δn_j, $j \neq 1, 2$, must vanish if Eq. (13) is to hold. We must have therefore

$$\log (g_j \varrho_j \varDelta z) + \nu - \lambda z_j - \log n_j = 0$$

$$n_j = g_j \varrho_j \, e^{(\nu - \lambda z_j)} \, \varDelta z \tag{14}$$

Eq. (14) is the end result of our statistical considerations and is equivalent to

$$f(z) = g(z) \, \varrho(z) \, e^{(\nu - \lambda z)} \tag{15}$$

We note that ν can be eliminated through the normalizing condition

$$\int_0^\infty f(z) \mathrm{d}z = N \tag{16}$$

which is equivalent to Eq. (2) and that λ can then be expressed in terms of the average cost of transportation Z/N through

$$\bar{z} = \frac{Z}{N} = \frac{\displaystyle\int_0^\infty z f(z) \mathrm{d}z}{\displaystyle\int_0^\infty f(z) \mathrm{d}z} \tag{17}$$

The quantity λ is in fact the inverse of the average cost of transportation $\bar{z}$ apart from a numerical factor which depends on the detailed form of $g(z)$ and $\varrho(z)$.

The form of the distribution function $f(z)$, Eq. (15), obtained by the foregoing arguments resembles closely that proposed as a model by WARDROP[1]. The salient feature of his model, the exponential factor $e^{-\lambda z}$, is reproduced by our treatment wherein it is a direct consequence of the fixed total amount Z spent on home-to-work-and-return trips per day by the population. Further WARDROP formulates

References p. 84

his model in such a way that the density of household sites, $\varrho(z)$, is not included in the distribution factor. We therefore interpret the factor z^{-1} in WARDROP's distribution function as our preference factor $g(z)$. Inasmuch as WARDROP has fitted his model directly to the statistical information available, we can regard z^{-1} as the experimentally determined dependence of $g(z)$ on z, a reasonable result.

This agreement with WARDROP's model permits us to say with some justification that we have developed a successful theory of the relative distribution of households and places of work. Nevertheless, it is only a very modest success. There are some very serious oversimplifications in our treatment. We have neglected any correlation or interaction between different places of employment, or between different households. Further, we have, so to speak, averaged over all factors determining the distribution of population within a given area other than the cost of transportation. One such factor, the cost or rent of the household may be at least as important as the cost of transportation and may not be independent of it. We have not been very explicit in our discussion of the preference factor $g(z)$ or of the density of household sites, $\varrho(z)$.

Agreement with WARDROP's model, however, was not our principal goal in developing the above treatment. Rather we wish to call attention to the general applicability of the methods of statistical mechanics to a wide class of problems of a primarily economic, social, or demographic character. In this we hope we have succeeded despite the obvious oversimplifications and inadequacies of our arguments.

Indeed, attempts to improve the arguments are even more suggestive of the applicability of statistical mechanics to such problems. For example, there was no restriction placed on the number of workers n_j with households in the jth range of z. We allowed distributions of workers in which n_j exceeded the number of household sites, $\varrho_j \Delta z$. We used the so-called 'Boltzmann statistics'[2], which does not apply to cases such as ours where a household site may be unoccupied, be occupied by one worker, or, infrequently, by more than one worker, a situation we ignore. Instead, we should have used the so-called 'Fermi-Dirac statistics'[2], which holds when sites are either unoccupied or singly occupied. The derivation of the distribution function for the Fermi-Dirac case is sufficiently similar to that for the Boltzmann case for its explicit derivation to be unprofitable here; the result is

$$f(z) = \varrho(z)F(z) \tag{18}$$

where $F(z)$, the Fermi factor

$$F(z) = \frac{1}{g(z)^{-1} \, e^{\lambda(z-\mu)} + 1} \tag{19}$$

gives the probability that a given household site is occupied. The parameters μ

References p. 84

and λ are again determined by the total number of workers N and the average cost of transportation $\bar{z}$ through Eq. (16) and (17), respectively.

The parameter μ is analogous to the chemical potential in thermodynamic problems in physics, and λ^{-1} is analogous to temperature. These analogies can be usefully exploited. When N tends to be small or the average spent on transportation $\bar{z}$ large, $\lambda\mu$ becomes large and negative. The Fermi factor is well approximated by an exponential

$$F(z) \simeq g(z)\, e^{\lambda(\mu-z)} \qquad \lambda\mu \ll 0 \tag{20}$$

and we find a very diffuse population distribution. When, on the other hand, N tends to be large or $\bar{z}$ small, μ becomes large and positive, the Fermi factor

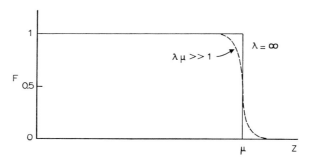

Probability of occupancy of a household site $F(z)$ as a function of the cost of transportation z from home to work and return per day. At most one worker is allowed to occupy a household site. When the average cost $\bar{z}$ the population allows for transportation goes to zero (or λ goes to ∞), the distribution of population is compact and has a sharp edge (solid line). When $\bar{z}$ is allowed to increase slightly ($\lambda\mu \gg 1$), a 'suburban' tail develops beyond the city limits at $z = \mu$.

approaches the step function as shown in the figure, and we find a compact distribution of population. The dotted curve in the figure clearly describes a modern city well. In the interior of the city ($z < \mu$), all household sites are occupied ($F(z) = 1$). The 'city limit' is characterized by a cost of transportation $z = \mu$. When z becomes larger than μ, we enter the 'suburban' tail of the distribution, where $F(z)$ may again be approximated by an exponential. Thus WARDROP's model is only strictly applicable to workers who live outside the city limits.

REFERENCES

1 J. G. WARDROP, in R. HERMAN (ed.), *Proc. Symposium on Traffic Flow, Detroit 1959*, Elsevier, Amsterdam, 1961, p. 57.
2 R. B. LINDSAY, *Physical Statistics*, Wiley, New York, 1941.

Multicopy Traffic Network Models*

A. CHARNES AND W. W. COOPER

Northwestern University, Evanston, Illinois, and Carnegie Institute of Technology, Pittsburgh, Pennsylvania

ABSTRACT

This paper is concerned with mathematical models for predicting gross or over-all traffic flow in contrast to other theories which try to predict behavior of individual vehicles (or individual vehicles in groups) on a single artery, or which are concerned with properties of traffic flow at a single intersection. The models achieve analytical formulations for qualitative principles enunciated by WARDROP. They involve different interpretations of WARDROP's suggestions than those of PRAGER whose analytical formulations were possible only by imposing conditions which are too restrictive to encompass many traffic prediction problems. In particular, WARDROP's second principle, "All alternate routes which are actually used by vehicles travelling between two points require the same journey time by an individual vehicle and less time than on an unused route", is interpreted and amplified to mean same journey time for vehicles travelling from a given origin to a given destination.

The models presented range from multi-person game theoretic models to linear programming models. The tremendous advantages of the latter type for study of changes in arterial characteristics and for quantitative evaluation of effects of arterial policy (*e.g.* political or land developmental considerations) are indicated.

INTRODUCTION

Rational long-range planning of city traffic networks has focussed attention on the desirability of research directed toward an expedient method for developing and analyzing traffic loads and requirements from origins to destinations over city streets. Empirical approaches, including simulation with the aid of high speed computers, have tended to bog down because of the scale and complexity of these problems. Until recently, conceptual or analytical principles applying to whole systems in ways which can be used for guidance have been lacking.

* This research was supported in part by Nonr-1228 (10), Project 047-021 at Northwestern University and in part by Nonr-760, Project NR-047011 at Carnegie Institute of Technology. Reproduction of these materials in whole or in part is permitted for any purpose of the U. S. Government.

References p. 96

Attempts have been made to respond to this situation. For example, WARDROP[11] enunciated two rational (qualitative) principles which might plausibly be applied to the movement of traffic in a network when the critical controlling conditions can be viewed as allowing sufficient repetitions for the emergence of 'steady-state' patterns in the flows. Two distinct examples which often are regarded as fulfilling such conditions are (1) rush-hour and (2) average daily flows of traffic. Traffic engineers have attempted to apply principles such as these in a variety of ways. With given origin–destination requirements, for example, they have attempted to modify the selection of routes from origins to destinations in a manner which would yield traffic patterns more closely in accord with such principles. Because of the tremendous number of routes which are possible from even a single origin to a single destination, any attempt to implement these principles by direct route selection and examination is virtually impossible.

To aid in solving this kind of problem, PRAGER[9] formulated a mathematical model of traffic-network flows which would, by means of an extremal principle, imply satisfaction of WARDROP's second principle *viz.* all alternate routes which are actually used by vehicles travelling between two points require the same journey time by an individual vehicle and less time than on an unused route.

As PRAGER notes, his analytical formulation allows for satisfying WARDROP's principle by associating a journey-time potential at each node of the traffic network in a fashion which is analogous to the association of electrical potentials (voltages) at the nodes of a direct-current electrical circuit. To implement this analogy, he introduced the notions of main- and cross-stream directions for the traffic on any branch, the directions of each being prescribed in advance. Moreover, only the overall origin effluxes and the destination influxes could be prescribed along with the conservations of flows at nodes; also, there was to be no transfer between main and cross-stream flow, and the latter were to be small relative to the former.

Thus, although PRAGER achieved an analytic formulation he left unresolved the following difficulties: (1) the problem of *a priori* specification of main- and cross-stream directions, and (2) his approach did not assure that specific origin-to-destination requirements would always be met. An alternate approach will therefore be explored here in a way which (a) shifts the burden of predicting main- and cross-stream directions onto the model, (b) allows main- and cross-stream flows to assume any relative orders of magnitude, (c) permits interchange between main- and cross-stream directions in any manner which is consistent with the network, (the notion of main- and cross-streams are not required in the formulation which will be used) and, (d) ensures that all origin-to-destination requirements will be met.

SIMULATION BY MEANS OF EXTREMAL PRINCIPLES

Simulation and optimization are sometimes regarded as separate, almost contrasting, approaches. Confusion may therefore be avoided if it is born in mind that the objective here is simulation. Extremal methods are used to secure simplifications (and thus reduce the burden of analysis) while ensuring that prescribed principles (*e.g.* WARDROP's second principle) are preserved along with other conditions (*e.g.* origin-to-destination requirements) which need to be considered. By use of these methods (as will be shown) it is possible to avoid tedious and unwieldy comparisons of nonlinear functions of the traffic flows over alternate routes, of the kind used, say, in the engineering studies described in the preceding section.

Principles such as 'least action', 'least constraint', 'minimum energy dissipation', etc., have played this kind of simplifying role for the analysis of problems in physics. The extremal principles to be utilized here may be regarded in an analogous fashion. To state the matter differently, the idea which motivates this investigation is that by means of suitable extremal principles it should be possible to obtain a simple approach which yields the desired characterizations[12]. This includes conformance with principles which are believed to hold as well as other conditions which might be specified. In addition it should be possible to orient this kind of analytical approach so that additional insight may be attained from which new modes of attack can be devised for these (and related) kinds of problems. One by-product of the approach to be used here, for example, is that it makes analytical evaluations possible for planning changes in traffic networks and predicting flows.

POLYEXTREMAL CHARACTERIZATIONS OF A SINGLE NETWORK

Traffic patterns formed from 'emergent (repetitive) steady-state' conditions of the kind which are of interest here can be viewed as resulting from a 'shaking down' (toward equilibrium) which occurs through the coursing of many vehicles over the network in pursuit of their objectives. This suggests a game-theoretic analogy in which a player is associated with each origin who seeks to attain distribution of the vehicles at that origin in the required amounts at various destinations. 'Artificial' or 'phony' origins may be employed when different types of vehicles (which congest) need to be considered.

DUFFIN[5] (and, following him, D'AURIAC[4] and BIRKHOFF AND DIAZ[1]) has shown how specific laws of resistance to current flow in branches of a network may be directly associated with an extremal principle which generates the flow in a non-leaky network. Moreover, prescription of a network resistance law for each branch and conservation of current conditions at nodes implies the existence of a potential defined at the nodes (see BIRKHOFF AND DIAZ[1]). It is possible therefore to determine the potential principle which holds for a network (of direct-current type) when the resistance laws for current branches are prescribed.

Consider, once more, the game theory analogy which associates a player with each origin. Each such player strives to fulfill his flow requirements while operating in a network with resistance characteristics (*e.g.* time of flow over branches) which are partly determined by the flow actions of other players. The resistance laws are therefore completely specified only when the other players' actions are specified.

If it is assumed that each player utilizes paths in accordance with a potential principle like WARDROP's (so that the totality of traffic flow will conform to such a principle), then, by DUFFIN's construction, it is possible to consider this player as striving to minimize the sum of the integrals of his branch-current resistances.

A multiperson game situation thus emerges in which each player is striving toward a specific objective which he only partially controls. A plausible concept which may be applied to situations of the kind which are of interest here is the idea of a 'NASH[8] equilibrium' borrowed from the theory of games[10]. This equilibrium has the property that each player's specification of traffic flows is such that if he attempts any deviation from these routes he will worsen his value for the criterion function. Stated differently, the player controlling the vehicles at this origin will under such attempted deviations also depart, individually, from satisfying the WARDROP principle.

This concept suggests iterative approaches which are easy to apply in determining such an equilibrium point[12]. For example, the actual network may be 'loaded' with each player's requirements in a way which optimizes the resistance functions as determined by the previous loadings. After all loadings have been made the process can be repeated with the initial player optimizing on the resistance functions which result from the loadings assigned to the other players. When this process of loading and reloading converges, the result is a NASH equilibrium point.

Normal mathematical arguments and formulae will not be introduced until the next section where a more advantageous view (and notation) can be secured which can be transcribed, if desired, into the development of the present section. It may be noted here, however, that the existence of such equilibrium points can be established mathematically by fixed point considerations such as those employed by NASH[8] in his original development. Actually, at least in small examples, there is no difficulty in resolving questions of existence. Indeed, it is sometimes possible to exhibit numerous equilibrium points. The interpretation of such (multiple) points of equilibrium is still open for investigation, but it should be recognized that the NASH equilibrium point is, in general, stable only against deviations of a single player. Requirements for stability against multiplayer deviations would, of course, narrow the number of possibilities.* Furthermore, randomization processes are available which can be used to single out a unique equilibrium point if it is deemed expedient to avoid questions associated with non-uniqueness of solutions.

MULTICOPY NETWORK MODELS

The polyextremal characterization on a single network discussed in the last section offers one approach and suggests another. It is possible to comprehend the previous designs within the framework of a single extremization in a way which opens up all of the evaluative possibilities and computational efficiencies of mathematical (and even linear) programming.

A copy of the network may be associated with each origin in the following manner. With the αth origin is associated the current x_j^α in the jth branch and an orientation (the same for all copies of the network) is assigned to each branch[3]. The Kirchhoff node conditions (conservation of current) for the αth copy are then

$$\sum_j \varepsilon_{ij} x_j^\alpha = E_i^\alpha \tag{1}$$

where ε_{ij} are the incidence numbers[1], and E_i^α are the effluxes or influxes associated with the αth origin-to-destination requirements.

The flow from a single origin to multiple destinations under minimization for an additively separated convex function of the currents will result in a single direction of flow for the αth copy in any particular branch. This flow may be in the same direction or opposite to the branch orientation. The x_j^α may therefore be left unrestricted and their sign used to determine the direction of flow for the branch current in the αth copy. For the actual road network the resistance function for the actual branch will depend on the total volume of traffic, without regard to direction, in that branch, *i.e.* it is a function of the total congestion in that branch. The term 'actual branch' is employed to distinguish branches of the network from others which may be artificially introduced for a variety of reasons, as will be seen, to simplify the analysis.

The integrated resistance functions, R_j, of the actual network will be assumed continuous, convex and non-decreasing. They can thus be written

$$R_j \left(\sum_\alpha |x_j^\alpha| \right) \tag{2}$$

where the vertical strokes indicate that absolute values of these variables are being considered.

---

* Considerations of these kinds, called 'k-stability,' may perhaps more usefully be considered when alterations in the network are being contemplated.

The single extremal principle is

$$\min \{\textstyle\sum_j R_j \,(\textstyle\sum_\alpha |x_j{}^\alpha|)\}$$

subject to   (3)

$$\textstyle\sum_j \varepsilon_{ij} x_j{}^\alpha = E_i{}^\alpha .$$

This can, in general, be written as the single expression

$$\min \{\textstyle\sum_j R_j \,(\textstyle\sum_\alpha |x_j{}^\alpha|) + \textstyle\sum_\alpha \textstyle\sum_i \lambda_i{}^\alpha \,[\textstyle\sum_j \varepsilon_{ij} x_j{}^\alpha - E_i{}^\alpha]\} \qquad (4)$$

on introducing the Lagrangean multipliers $\lambda_i{}^\alpha$. As is well known, the $\lambda_i{}^\alpha$, at an optimum, play the role of potentials. This remark and the expression (4) may serve to justify the opening observations in this section.

The formulation (3) yields a convex programming problem. The necessary and sufficient conditions for an optimum in such problems established by KUHN AND TUCKER[7], show that it can stand the presence of any finite number of additional linear inequalities on the $x_j{}^\alpha$. Differentiability of the R_j is not required. Finally, the specific type of resistance function (2) can be replaced by other anisotropic resistance functions without losing the advantages of this formulation.

Particular attention may profitably be given to the case in which the R_j are piecewise linear. The principle can then be written in the form

$$\min \{\textstyle\sum_\alpha \textstyle\sum_j R_j |x_j{}^\alpha|\}$$

subject to

$$\textstyle\sum_j \varepsilon_{ij}\, x_j{}^\alpha = E_i{}^\alpha \qquad (5)$$

$$\textstyle\sum_\alpha |x_j{}^\alpha| \leq \varDelta_j$$

corresponding to replacement of the original jth branch by several new branches, each with a definite capacity and a different integrated resistance law which is linear in the total traffic on it. (The generic designation j is applied to the new branches as well as the old ones for economy in notation.) The capacity conditions are stated in the second set of constraints.

This second set of constraints can be replaced by an equivalent set which is linear in the $x_j{}^\alpha$. This would require 2^α explicit linear constraints for each branch, j. Also, as is well known[2,3], the absolute value type of functional (5) can be replaced (for extremization) by one which is linear. This is done by introducing new non-negative variables in a way which yields an equivalent linear programming problem.

References p. 96

To justify the remaining remarks in the opening paragraph of this section this equivalent problem may now be stated. In shorthand fashion (which yields a superficially linear appearance) this problem is

Find
$$\min \left\{ \sum_j R_j \left(x_{j+}{}^\alpha + x_{j-}{}^\alpha \right) \right\}$$

Subject to
$$\sum_j \varepsilon_{ij}{}^\alpha \left(x_{j+}{}^\alpha - x_{j-}{}^\alpha \right) = E_i{}^\alpha$$

$$-\sum_\alpha \left(x_{j+}{}^\alpha + x_{j-}{}^\alpha \right) \geq -\varDelta_j$$ (6)

$$x_{j+}{}^\alpha, \; x_{j-}{}^\alpha \geq 0 \, .$$

The dual to this problem may, similarly, be stated as

Find
$$\max \left\{ \sum_\alpha \sum_i \phi_i{}^\alpha E_i{}^\alpha - \sum_j \psi_j \varDelta_j \right\}$$

subject to
$$\left| \sum_\alpha \phi_i{}^\alpha \varepsilon_{ij}{}^\alpha - \psi_j \right| \leq R_j$$

$$\psi_j \geq 0$$

Optimal solutions $x_j{}^{*\alpha}$ to the direct (minimizing) problem and $\varphi_i{}^{*\alpha} \, \psi_j{}^*$ to its dual have the property

$$\sum_j R_j \, \mathrm{sgn} \, \left(x_j{}^{*\alpha} \right) x_j{}^{*\alpha} = \sum_\alpha \sum_i \phi_i{}^{*\alpha} E_i{}^\alpha - \sum_j \psi_j{}^* \varDelta_j \, ,$$

where $\mathrm{sgn} \, (z) = \begin{cases} + 1, z > 0 \\ 0, z = 0 \\ - 1, z < 0 \end{cases}$

The $\varphi_i{}^{*\alpha}$ and $\psi_j{}^*$ provide evaluators for the rates of changes in $E_i{}^\alpha$ (effluxes or influxes for the αth copy) and $\varDelta_j$ (the critical points at which changes in 'resistance' to traffic volumes occur). Hence a means is provided for evaluating changes associated with these features of the network and for assessing resulting traffic volumes and times.

The efficiency of linear programming methods of solution has already been observed. Additional gains may be secured by computation devices directed towards taking advantage of the structural features[2] of the model which emerge from this formulation. Notice that the various copies of the network are bound (or hooked) together only through the '$\varDelta_j$ conditions' in Eqs. (6). Every basis for

References p. 96

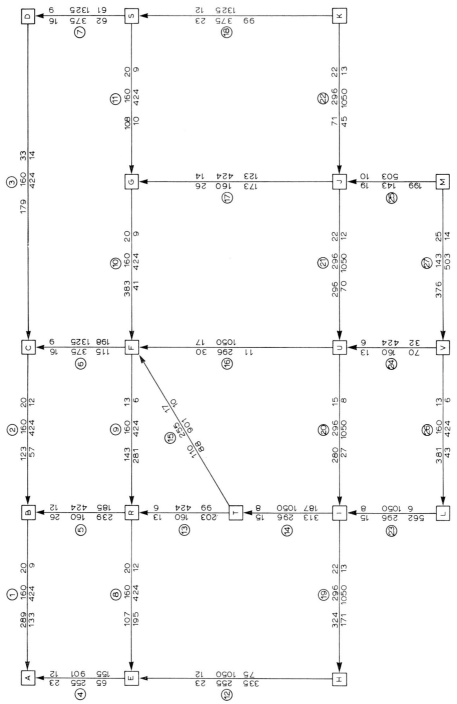

Fig. 1. Characteristics of arterial network and optimal loads.

the total problem is composed of a basis for each copy of the network (ignoring the hook-up conditions) plus one more vector for each j. Node potentials for every copy of the network, correlated only through the ψ_i, are thus made available and the possibility of special linear programming methods involving parallel computation on the copies indexed by α thereby opened. It might be further noted that the model can be viewed as an instance of a dynamic network programming problem in which the αth copy corresponds to the conditions in the αth time period and the j conditions provide inter-period connections. This opens access to further generalizations which will be dealt with in subsequent papers.

AN EXAMPLE OF A MULTICOPY TRAFFIC NETWORK FLOW

To better illustrate and fix the ideas we present an example computed* from a modification of data pertaining to a small town in Indiana supplied by Dr.

TABLE I

ORIGIN AND DESTINATION REQUIREMENTS

Node	\multicolumn{11}{c}{Copy}										
	1	2	3	4	5	6	7	8	9	10	11
A	50	14	22	24	38	5	35	42	45	−209	
B	35	11	32	18	27	6	34		34	−233	
C			2	62	20	−54					
D	10	27	−215								
E	55	16	24	27	71	9	46	43	52	103	−184
F	60	19	29	31	45	10	84	54	86	57	121
G	40	12	30	20	26	7	−329				
H	170	7	12	22	−318						
I	55	18	27	29	64	9	46	120	50	49	63
J	55	14	31	23	27	8	84	−293			
K	45	−170									
L		32	6	−256							
M	−575										
R											
S											
T											
U											
V											

Negative sign denotes an influx

D. CARROLL, Mr. P. CASWELL and Miss E. L. GARDNER of C.A.T.S. (Chicago Area Transportation Study).

* We wish to thank Mr. A. BEN-ISREAL, The Technological Institute, Northwestern University, for the computation of and computational experimentation (involving multicopy methods and mixing routines of CHARNES, COOPER AND LEMKE) in connection with this example.

We employ the linear programming type of model as discussed above. Each copy has 18 nodes (indicated by capital letters) and 27 branches (numbered 1 through 27) as indicated in Fig. 1. Origin and destination requirements are listed in Table I. On each branch in Fig. 1, the loading (in each direction) prescribed by

TABLE II

BRANCH FLOWS BY COPIES

Branch	\multicolumn{11}{c}{Copy}
	1	2	3	4	5	6	7	8	9	10	11
1	50	14	22	24	$\overline{27}$	5	35	42	97	$\overline{106}$	
2		25	54			11	33			57	
3		25	154								
4					65				$\overline{52}$	$\overline{103}$	
5	85			42			36	76	$\overline{136}$	49	
6			$\overline{98}$	62	20	$\overline{43}$	33			57	
7	10	52	$\overline{61}$								
8		16	36		$\overline{74}$	9	46				$\overline{121}$
9		16	36		$\overline{74}$	$\overline{9}$	82		$\overline{86}$		$\overline{121}$
10	49	35			$\overline{26}$	$\overline{15}$	245	54			
11	$\overline{10}$	47	61								
12	55		$\overline{12}$	27	210			43			$\overline{63}$
13	85			42				76	$\overline{50}$	49	
14	85		$\overline{33}$	135	17	$\overline{9}$	46	76	50	49	
15			$\overline{33}$	93	17	$\overline{9}$	46				
16	11										
17	99		$\overline{31}$	20		$\overline{8}$	$\overline{84}$	54			
18		99									
19	225	7		49	$\overline{108}$			43			$\overline{63}$
20	16	25			$\overline{27}$			239			
21		57		$\overline{43}$	$\overline{27}$			239			
22	$\overline{45}$	71									
23	349		$\overline{6}$	213							
24	27	$\overline{32}$		43							
25	199										
26	349	32		$\overline{43}$							
27	376										

A bar denotes flow direction opposite to branch arrow in Fig. 1.

the extremal principle and the transit-time characteristics of the branch are presented in a three-column two-row array.

The left-hand column contains the vehicular traffic; the top row being reserved for that in the direction of the brand arrow, the bottom row being reserved for traffic in the opposite direction. For greater convenience these are also listed in detail, by copy, in Table II. The right-hand column contains two transit times associated with piece-wise linear approximation of the nonlinear transit time to traffic volume relationship; the numbers in the middle column are the respective total numbers of vehicles which can traverse the branch at the transit time specified in the same row, *i.e.* they are the 'Δ_j' referred to in the linear programming model.

In this particular example no branch carried a traffic volume greater than the traffic capacity of its best transit time. This is due in part to the capacities available

TABLE III

NODE AND CAPACITY EVALUATORS

Node	Copy										
	1	*2*	*3*	*4*	*5*	*6*	*7*	*8*	*9*	*10*	*11*
A	65	56	35	45	24	21	43	57	9	0	12
B	56	47	26	36	33	12	34	48	0	9	21
C	46	35	14	35	40	0	22	36	12	21	28
D	44	21	0	56	54	14	18	32	26	35	42
E	53	53	42	33	12	28	32	46	21	12	0
F	37	34	23	26	31	9	13	27	21	30	19
G	24	21	18	38	44	22	0	14	34	43	32
H	41	46	54	21	0	40	44	34	33	24	12
I	28	33	41	8	13	27	31	20	28	37	25
J	10	13	32	24	33	36	14	0	48	57	45
K	23	0	21	37	46	35	21	13	47	56	53
L	20	37	49	0	21	35	38	24	36	45	33
M	0	23	42	20	41	46	24	10	56	65	53
R	42	41	30	22	24	16	20	34	14	23	12
S	35	12	9	47	53	23	9	23	35	44	41
T	36	41	33	16	21	19	23	28	20	29	18
U	20	25	40	12	21	26	26	12	36	45	33
V	14	31	46	6	27	32	32	18	42	51	39

$\psi_5 = 2$, $\psi_9 = 1$, $\psi_{10} = 4$, $\psi_{17} = 0$, $\psi_{26} = 0$.

on some branches and in part to the spread between the best and the 'congested' transit times on other branches. By means of the 'node evaluators' (*e.g.* the $\varphi_i^{*\alpha}$) and the 'capacity evaluators' (*e.g.* the ψ_j^*) listed in Table III, one can appraise the reduction in transit time necessary for an unused branch to be used by traffic from any particular origin, and also the reduction in total vehicle-hours occasioned by unit increase in the traffic capacity on particular branches. For instance, $\psi_9 = 1$, $\psi_5 = 2$, $\psi_{10} = 4$ means that unit increase in the respective capacities on branches 9, 5, and 10 will produce respective decreases of 1, 2 and 4 vehicle-hours in total vehicle-hours. Similarly, $\varphi_B^A - \varphi_C^A = 56 - 46 = 10 < 12$, the transit time on branch 2 (C to B) means that branch 2 will only be used by traffic from origin A if the transit time is reduced from 12 to 10.

References p. 96

REFERENCES

1 G. BIRKHOFF AND J. B. DIAZ, Non-linear Network Problems, *Quart. Appl. Math.*, *13* (1956) 431–443.

2 A. CHARNES AND W. W. COOPER, Management Models and Industrial Applications of Linear Programming, *Management Sci.*, *4* (1957) no. 1.

3 A. CHARNES AND W. W. COOPER, *Non-Linear Network Flows and Convex Programming Over Incidence Matrices*, Naval Research Logistics Quart., *5* (1958) 231–240.

4 A. D'AURIAC, A propos de l'unicité de solution dans les problèmes des réseaux maillés, *La houille blanche*, *2* (1947) 209–11.

5 R. J. DUFFIN, Non-linear Networks, *Bull. Am. Math. Soc.*, *52* (1946) 833–838 and *53* (1947) 963–971.

6 D. GALE, A Theory of N-Person Games With Perfect Information, *Proc. Natl. Acad. Sci. U. S.*, *39* (1953) 496–501.

7 H. W. KUHN AND A. W. TUCKER, Non-linear Programming, in J. NEYMAN (ed.), *Proc. Second Berkeley Symposium on Math. Statist. Probability*, University of California Press, Berkeley, 1951.

8 J. F. NASH, Non-Cooperative Games, *Ann. Math.*, *54* (Sept. 1951).

9 W. PRAGER, Problems of Traffic and Transportation, *Proc. Symposium on Operat. Research Business Ind.*, Midwest Research Inst., Kansas City, Mo., 1954, p. 105–113.

10 J. VON NEUMANN AND O. MORGENSTERN, *Theory of Games and Economic Behavior*, Princeton Univ. Press, Princeton, 3rd Ed., 1953.

11 J. G. WARDROP, Some Theoretical Aspects of Road Traffic Research, *Proc. Inst. Civil Engrs.*, *1* (1952) 325–378.

12 A. CHARNES AND W. W. COOPER, Extremal Principles for Simulating Traffic Flow in a Network, *Proc. Natl. Acad. Sci. U. S.*, *44* (1958) 201–204.

On the Design of Communication and Transportation Networks

W. PRAGER

Brown University, Providence, Rhode Island

ABSTRACT

The morning rush hour traffic from the suburbs via streets and expressways to the central business district of a city has many features in common with the flow of telephone messages from the individual subscribers via feeders and trunks to a central telephone office. In view of this analogy, the paper briefly reviews work on the economic design of simple telephone networks and indicates to what extent techniques developed in this field may be applied to problems concerning the economic design of a network of streets and expressways.

INTRODUCTION

This paper is primarily concerned with the economic design of simple telephone networks. It is based on a report which I wrote in January 1958 for the Bell Telephone Laboratories, to which I am indebted for the permission to present this material at this symposium. In briefly describing this work, I shall endeavour to indicate how similar techniques may be applied to problems concerning the economic design of a network of streets and expressways.

The city for which a telephone network is to be designed is divided into a central region and a number of peripheral regions. Each of the latter is connected by a *trunk* to a central office, which is situated in the central region. The individual telephone subscriber is connected by a *feeder* either to the trunk head of his region, or directly to the central office if he lives in the central region. Thus, a call within the city in general proceeds by feeder to the trunk head of the originator's region, then by trunk to the central office and again by trunk to the trunk head of the recipient's region, and finally by feeder to the recipient. One of the trunk links will disappear from this pattern, if the call either originates or terminates in the central region, and the call will not use any trunk at all if it orginates and terminates in this region.

Feeders and trunks have to follow the city streets which run, say, north–south and east–west. Since the dimensions of a typical city block are supposed to be

small in comparison with the over-all dimensions of the city, the analysis will be
simplified by treating the streets as continuously distributed.

The cost of providing capacity for certain peak traffic conditions does not
depend on the direction in which the individual messages flow. It is therefore
convenient to visualize the entire traffic as flowing from the subscribers to the
central office, regardless of whether the individual subscriber originates or receives
the message under consideration. Viewed in this manner, this telephone traffic has
some similarity with the morning rush hour traffic of employees to the central
business district, the feeders corresponding to city streets and the trunks to
expressways from the suburbs to the city center. As the telephone trunks, these
expressways would have to run north–south or east–west, but expressways of
other directions can be treated by an obvious modification of the method described
below.

The principal difference between the two traffic patterns is this: all telephone
traffic must go through the central office, whereas the road traffic to the central
business district is perturbed by traffic from suburban residences to suburban
places of employment, which by-passes the central business district. Since, however, the volume of this perturbing traffic is small compared to the volume of the
main traffic to the city center, the study of the simpler pattern of telephone traffic
may nevertheless give useful pointers concerning the more complicated pattern
of road traffic.

Resuming the discussion of the telephone network, we note that, dropping the
distinction between traffic originating in a given area and traffic terminating there,
we may henceforth speak of the traffic in this area. Letting the x and y axes point
to the east and north, respectively, we denote by dA the element of area containing
the generic point x, y and by $dQ = f(x, y)\, dA$ the traffic in dA. The function
$f(x, y)$ will be called the load distribution on the network; its value at a specific
point x, y will be called the load at this point.

In discussing the optimal design for a given load distribution $f(x, y)$, it is often
convenient to embed this distribution in the one-parameter family $\lambda f(x, y)$ of
load distributions. The positive parameter λ that singles out a specific member
of the family will be called the load factor.

The cost of providing capacity for the traffic Q in a trunk or feeder is assumed
to be

$$c_T = \alpha + \beta Q \tag{1}$$

per unit length of a trunk and

$$c_F = bQ \tag{2}$$

per unit length of a feeder; here α, β, and b are constants and $b > \beta$. A network
design is called optimal for given loads and given values of the constants α, β

and b, if the total cost of providing the necessary feeder and trunk capacities is not larger than for any other network that is capable of handling the given loads. (The expression "not larger" is preferable to "smaller" because there may not be a unique optimal design.)

LOW LOADS

Over the years, the loads on a telephone network will increase and their spatial distribution may change. For the sake of mathematical simplicity, the spatial distribution $f(x, y)$ will however be treated as constant in the following, and the optimum design of the network will be discussed in dependence on the load factor λ. For given $f(x, y)$ and given values of the constants α, β and b, there exists, in general, a critical value λ^* such that the optimal design does not involve any trunks as long as $\lambda < \lambda^*$.

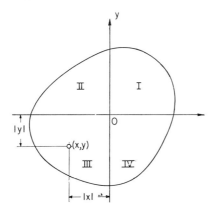

Fig. 1. Optimum position of single office O.

To derive a criterion for the optimal location of the central office in this case, let the coordinate origin O coincide with the tentative position of the central office (Fig. 1). The intensity Q_I of the traffic in the first quadrant is

$$Q_I = \lambda \int f(x,y) \, \mathrm{d}A \tag{3}$$

where the integration is extended over the first quadrant. Similar expressions may be written for the traffic in the other quadrants.

The feeder from the generic point x, y to the central office at the origin follows the streets and therefore has the length $|x| + |y|$, which will be called the 'Manhattan distance' of the point x, y from the origin. When the office is given an infinitesimal eastward displacement $\mathrm{d}x$, the Manhattan distance of any subscriber in quadrants I and IV from the office is decreased by $\mathrm{d}x$ and that of any subscriber in quadrants II and III is increased by the same amount. The corresponding change in cost is $b(Q_{II} + Q_{III} - Q_I - Q_{IV})$, and this quantity must vanish if the office has the best possible location. A similar condition may be derived from the consideration of a northward displacement of the office. We therefore have the following result: for the optimal location of the central office, the traffic to the east of the office equals the traffic to the west of the office, and this statement remains valid when the words 'east' and 'west' are replaced by the words 'north' and 'south'.

Note that this criterion uniquely determines the optimal location of the office. It follows from this criterion that the traffic in quadrant I must equal that in quadrant III, and the traffic in quadrant II must equal that in quadrant IV. These results can be obtained directly by considering infinitesimal displacements of the central office along the bisectors of the quadrants instead of displacements along the coordinate axes.

HIGH LOADS

When the load factor λ exceeds the critical value λ^* introduced in the previous section, the optimal design involves trunks. Let us first assume that the positions of the central office and the trunk heads are dictated by other considerations, and that only the boundaries of central and peripheral regions are to be discussed from the economic point of view.

In Fig. 2, H_1 and H_2 are trunk heads of neighboring peripheral regions and O is the central office. The boundary b_{12} between the regions served by the trunk heads H_1 and H_2 must be so drawn that the cost of providing capacity for additional traffic at the typical point P_{12} of b_{12} does not depend on whether this traffic its routed via H_1 or H_2. Let r_1 and r_2 be the Manhattan distances of P_{12} from H_1 and H_2, respectively, and R_1 and R_2 the Manhattan distance of H_1 and H_2 from O. The optimum principle formulated above then furnishes the equation

$$br_1 + \beta R_1 = br_2 + \beta R_2$$

or

$$r_1 - r_2 = \frac{\beta}{b}(R_2 - R_1) \tag{4}$$

Since the right-hand side of Eq. (4) does not depend on the choice of P_{12} on b_{12}, Eq. (4) characterizes b_{12} as the locus of points for which the difference of their Manhattan distances from H_1 and H_2 has a constant value. A line of this kind may be called a Manhattan hyperbola with the foci H_1 and H_2. Fig. 3 shows such hyperbolas.

The optimal boundary between the region served by the trunk head H_1 and the central region can be determined in a similar manner. The cost of providing capacity for additional traffic at the typical point P_{01} of b_{01} must not depend on

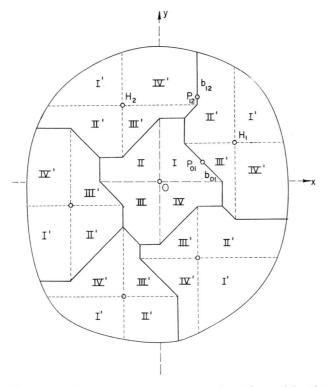

Fig. 2. Boundaries of regions served by one central and five peripheral offices.

whether this traffic is routed by feeder to H_1 and hence by trunk to O or directly by feeder to O. This optimum principle yields the equation

$$r_0 - r_1 = \frac{\beta}{b} R_1 \qquad (5)$$

where r_0 denotes the Manhattan distance of P_{01} from O. As is shown by Eq. (5), the boundary b_{01} is again a Manhattan hyperbola.

Next, let us investigate the optimal position of a trunk head within its region. Letting H_1 in Fig. 2 be the optimal position of the trunk head of the first region, study the change in cost caused by an infinitesimal displacement ds $\sqrt{2}$ of the trunk head along the bisector of the quadrant I' of the first region. The Manhattan distance of the trunk head from O or any subscriber in III' is thereby increased by 2ds and that from any subscriber in I′ is decreased by 2ds, whereas the Manhattan distance of the trunk head from the subscribers in II′ and IV′ remains the same. The necessary change in feeder capacity costs $2b(Q'_{III} - Q'_I)$ ds, where Q'_I is the traffic in I′ and Q'_{III} that in III′. If the total traffic in region I is denoted by Q', the necessary change in trunk capacity costs $2(\alpha + \beta Q')$ ds.

On account of the displacement of H_1, the boundaries of the first region have to be modified, if they are to be optimal for the new position of the trunk head. This may mean, for instance, that points of the first region adjacent to the original boundary b_{12} will be incorporated in the second region corresponding to the new position of H_1. This change, however, does not affect the cost of the network, because b_{12} has already been drawn in such a manner that the cost is independent of whether traffic from points on b_{12} is routed via H_1 or H_2. A similar reasoning applies to the other boundaries of the first region. The total change in cost caused

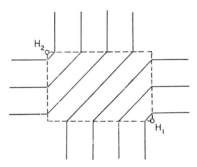

Fig. 3. Manhattan hyperbolas as possible boundaries of regions served by offices H_1 and H_2.

by the considered displacement of H_1 is therefore the sum of the two terms derived above. For optimality this must vanish, yielding the condition

$$\alpha + \beta Q' + b(Q'_{III} - Q'_I) = 0 \tag{6}$$

The same condition holds for the trunk heads of the other peripheral regions, provided the quadrants are labelled in a counter-clockwise manner so that a displacement of the trunk head along the bisector of quadrant I′ increases the Manhattan distance between trunk head and central office (Fig. 2).

A similar discussion of a displacement of the trunk head along the bisector of quadrant II′, yields the optimality condition

$$Q'_{II} = Q'_{IV} \tag{7}$$

where Q'_{II} is the traffic in quadrant II′ and Q'_{IV} that in quadrant IV′.

We finally investigate the optimal position of the central office O. By shifting it by ds $\sqrt{2}$ along the bisector of quadrant I in Fig. 2, we increase the cost of feeder capacity in the central region by $2(Q_{III} - Q_I)\,ds$, where, for instance, Q_I is the traffic in quadrant I of the central region. The considered shift also affects the lengths of some trunks, but may not alter the lengths of others. Let Q_+ be the total traffic in the n_+ regions whose trunks are lengthened by $2ds$ and Q_- the total traffic in the n_- regions whose trunks are shortened by the same amount. The cost of the necessary change in trunk capacity is then $2[\alpha(n_+ - n_-) + \beta(Q_+ - Q_-)]ds$. Optimality therefore requires that

$$\alpha(n_+ - n_-) + \beta(Q_+ - Q_-) + b(Q_{III} - Q_I) = 0 \qquad (8)$$

A second optimality condition can be obtained by considering a shift of O along the bisector of quadrant II. It has the form

$$\alpha(n_+{}^* - n_-{}^*) + \beta(Q_+{}^* - Q_-{}^*) + b(Q_{IV} - Q_{II}) = 0 \qquad (9)$$

where, for instance, $Q_+{}^*$ is the total traffic in the $n_+{}^*$ regions whose trunks are

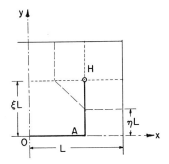

Fig. 4. Square area served by central office O and four peripheral offices H.

lengthened by this shift and Q_{II} is the traffic in quadrant II of the central region.

In the preceding discussion, it was tacitly assumed that the trunk head or the central office can be displaced in any direction. When the load distribution has an axis of symmetry, the central office will lie on this axis and only displacements along this axis need to be considered. This means that Eq. (8) and (9) are no longer valid individually, but must be combined to represent a displacement along the axis of symmetry. A similar remark applies to trunk heads on axes of symmetry.

EXAMPLE

Let the city be a square of the side $2L$, assume the load distribution to be uniform and denote the total load on the city network by Q. By symmetry, the central office is at the center of the square, and we shall study the network that has four trunk heads on the diagonal rays from the center. Fig. 4 shows one quadrant with tentative positions of the trunk head and the boundary of the region served by it. With the notations of Fig. 4, Eq. (5) yields $\eta = \beta \xi / b$. For the following analysis, it will be assumed that $\beta/b = 0.1$; accordingly

$$\eta = 0.1 \, \xi \tag{10}$$

By symmetry, the only admissible displacement of the trunk head is along the bisector of the quadrant. Since, however, all trunk heads must be displaced by equal amounts ds $\sqrt{2}$ at the same time, the total increase in trunk length for the entire city is 6 ds, because the trunk segment OA serves both the first and fourth quadrants. For the first quadrant, Eq. (6) must therefore be modified by replacing α by $3\alpha/4$. On account of the uniform distribution of the total traffic Q over the area $4L^2$ of the city, the traffic in, say, I' is obtained by multiplying the area

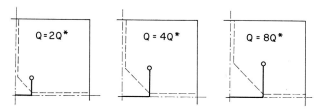

Fig. 5. Optimum location of peripheral office in problem of Fig. 4 (Q = actual load, Q^* = load at which peripheral offices first become economically rewarding).

$(1 - \xi)^2 \, L^2$ of I' by $Q/4L^2$. When the other traffic amounts appearing in the modified Eq. (6) are evaluated in this manner, the following equation is obtained

$$\frac{\alpha}{\beta Q} = 3.000 - 6.600 \, \xi + 1.695 \, \xi^2 \tag{11}$$

For a given value of $\alpha/(\beta Q)$ that root of Eq. (11) must be taken which satisfies $0 \leq \xi \leq 1$. Accordingly, no trunks should be provided when $Q \leq \alpha/(3\beta) = Q^*$. Fig. 5 shows the lay-outs for the total loads $2Q^*$, $4Q^*$, and $8Q^*$. As Q increases indefinitely, the trunk head asymptotically approaches the position specified by $\xi = 0.525$.

CONCLUDING REMARKS

The conditions established in the previous section only ensure a local optimum. For the city discussed in the example, for instance, the absolute optimum for a given load may involve trunk heads on the coordinate axes rather than the bisectors of the quadrants, or on both, axes and bisectors. This question can only be answered by comparing the costs of the local optima for the different topologies.

Dynamic Behavior of Traffic with a Nonlinear Spacing–Speed Relationship

E. KOMETANI AND T. SASAKI

Department of Civil Engineering, Kyoto University, Kyoto, Japan

ABSTRACT

So far we have investigated a case where the car spacing is expressed as a linear function of the speeds of the lead car and the following car. But the car spacing in actual traffic flow is often expressed by a quadratic relation of the speeds of both cars except in the case of uniform traffic. In this paper we have dealt with traffic flow in which the car spacing is expressed by a quadratic relation of speeds, and have investigated the effect of nonlinearity upon the steady motion of the following vehicle, upon the variation of car spacing and on the variation of the average spacing after the lead car has suffered a disturbance. We have also described the results of the car-following experiments performed.

INTRODUCTION

When the traffic volume increases and passing maneuvers become more and more difficult, it comes about that the behavior of the lead car controls that of the following cars 'perfectly'. What kinds of phenomena will appear in the group of following vehicles in this case as a consequence of the motion of a lead car?

To solve such a problem, we have considered previously Eq. (1) as the fundamental equation of traffic dynamics

$$V_{k+1}(t) = f[X_k(t-T) - X_{k+1}(t-T), V_k(t-T)] \qquad (1)$$

The motion of the following cars, and especially the problem of stability, have been discussed for a case when f is considered as a linear function[2,3], in which case the following relationship exists

$$X_k(t-T) - X_{k+1}(t-T) = \alpha V_k(t-T) + \beta V_{k+1}(t) + b_0 \qquad (2)$$

where $X_k(t)$ and $V_k(t)$ denote the position and speed of the kth vehicle at time t and $X_{k+1}(t)$ and $V_{k+1}(t)$ likewise the position and speed of the $(k+1)$th vehicle at time t. The quantity T is the reaction time of a driver, and α, β, b_0 are constants. The left hand side of Eq. (2) gives the car spacing at time $(t-T)$. We have defined

References p. 118

the traffic flow described by Eq. (2) as traffic with linear spacing. Eq. (2) is valid so long as the speed difference is relatively small. Whereas in a case when the speed varies widely the car spacing can be assumed to vary with the square of the speed of the lead car and that of the following car. (This assumption is closely related to the fact that the braking distance of each vehicle is proportional to the square of its speed). In the latter case Eq. (2) is replaced by the following

$$X_k(t-T) - X_{k+1}(t-T) = \alpha V_k^2(t-T) + \beta_1 V_{k+1}^2(t) + \beta V_{k+1}(t) + b_0 \qquad (3)$$

where α, β_1, β and b_0 are all proportionality constants. In this paper we intend to consider the characteristics of traffic flow which are expressed by Eq. (3). The reason why Eq. (3) was adopted for a fundamental equation of traffic dynamics is explained in the appendix.

THE MOTION OF THE FOLLOWING VEHICLES SUBJECTED TO A SINUSOIDAL DISTURBANCE

Differentiating Eq. (3) with respect to time we obtain

$$V_k(t-T) - V_{k+1}(t-T) = \alpha \frac{d}{dt}\{V_k^2(t-T)\} + \beta_1 \frac{d}{dt}\{V_{k+1}^2(t)\} + \beta \frac{d}{dt} V_{k+1}(t) \qquad (4)$$

Within a traffic stream which is proceeding at a uniform speed V_0, consider a case where the kth vehicle is subjected to a disturbance. If the motion of the kth vehicle is expressed by

$$V_k(t) = V_0 + Z(t) \qquad (5)$$

then the motion of the $(k+1)$th vehicle is expressed, in general, by

$$V_{k+1}(t) = V_0 + Y(t) \qquad (6)$$

Substituting Eq. (5) and (6) into Eq. (4) we obtain the following relation

$$V_0 + Y(t-T) + (2\beta_1 V_0 + \beta)\frac{d}{dt} Y(t) + \beta_1 \frac{d}{dt}\{Y^2(t)\}$$

$$= V_0 + Z(t-T) - 2\alpha V_0 \frac{d}{dt} Z(t-T) - \alpha \frac{d}{dt}\{Z^2(t-T)\} \qquad (7)$$

Therefore the following relationship must exist

$$Y(t-T) + (2\beta_1 V_0 + \beta)\frac{d}{dt} Y(t) + \beta_1 \frac{d}{dt}\{Y^2(t)\}$$

$$= Z(t-T) - 2\alpha V_0 \frac{d}{dt} Z(t-T) - \alpha \frac{d}{dt}\{Z^2(t-T)\} \qquad (8)$$

References p. 118

When the motion of the lead car is described by

$$Z(t) = S \sin \omega t \tag{9}$$

the speed fluctuation of the following car is expressed by

$$Y(t) = A_1 \sin(\omega t - \varphi_1) + A_2 \sin(2\omega t - \varphi_2) + A_3 \sin(3\omega t - \varphi_3) + \ldots \tag{10}$$

If the system is linear, Eq. (10) is given by the first term only, whereas if the system is nonlinear, Eq. (10) describes a nonharmonic oscillation.

In order to determine $A_1, \varphi_1, A_2, \varphi_2, A_3, \varphi_3, \ldots$ in Eq. (10), we substitute Eqs. (9) and (10) into Eq. (8), put $\omega t - \varphi_1 = 0$, and then equate the coefficients of $\sin\theta$, $\cos\theta$, $\sin 2\theta$, $\cos 2\theta$, $\sin 3\theta$, $\cos 3\theta$, $\ldots$ on the two sides of the equation. Thus we can generally have the same number of equations as unknowns and can determine the unknowns $A_1, \varphi_1, A_2, \varphi_2, \ldots$ As a matter of fact A_2 is smaller than A_1, and A_3 smaller than A_2, and thus the values of A become smaller successively. Therefore Eq. (10) is approximately correct even if A_2, A_4 were neglected. Now, neglecting $A_4, A_5, \ldots$ and equating the coefficients of $\sin\theta$, $\cos\theta$, $\sin 2\theta$, $\cos 2\theta$, $\sin 3\theta$ and $\cos 3\theta$, we have the following relations

$$A_1 \cos \omega T - A_1 A_2 \beta_1 \omega \cos(2\varphi_1 - \varphi_2) - A_2 A_3 \beta_1 \omega \cos(\varphi_1 + \varphi_2 - \varphi_3)$$
$$= S \cos(\varphi_1 - \omega T) + 2S\alpha\omega V_0 \sin(\varphi_1 - \omega T)$$

$$A_1 (2\beta_1 V_0 + \beta) \omega - A_1 \sin \omega T - A_1 A_2 \beta_1 \omega \sin(2\varphi_1 - \varphi_2)$$
$$- A_2 A_3 \beta_1 \omega \sin(\varphi_1 + \varphi_2 - \varphi_3) = S \sin(\varphi_1 - \omega T) - 2S\alpha\omega V_0 \cos(\varphi_1 - \omega T)$$

$$A_1^2 \beta_1 \omega + A_2 \cos(2\varphi_1 - \varphi_2 - 2\omega T) - 2A_2 (2\beta_1 V_0 + \beta)\omega \sin(2\varphi_1 - \varphi_2)$$
$$- 2A_1 A_3 \beta_1 \omega \cos(3\varphi_1 - \varphi_3) = - S^2 \alpha\omega \cos(2\varphi_1 - 2\omega T) \tag{11}$$

$$A_2 \sin(2\varphi_1 - \varphi_2 - 2\omega T) + 2A_2 (2\beta_1 V_0 + \beta)\omega \cos(2\varphi_1 - \varphi_2)$$
$$- 2A_1 A_3 \beta_1 \omega \sin(3\varphi_1 - \varphi_3) = - S^2 \alpha\omega \sin(2\varphi_1 - 2\omega T)$$

$$3A_1 A_2 \beta_1 \omega \cos(2\varphi_1 - \varphi_2) + A_3 \cos(3\varphi_1 - \varphi_3 - 3\omega T)$$
$$- 3A_3 (2\beta_1 V_0 + \beta)\omega \sin(3\varphi_1 - \varphi_3) = 0$$

$$3A_1 A_2 \beta_1 \omega \sin(2\varphi_1 - \varphi_2) + A_3 \sin(3\varphi_1 - \varphi_3 - 3\omega T)$$
$$+ 3A_3 (2\beta_1 V_0 + \beta)\omega \cos(3\varphi_1 - \varphi_3) = 0$$

Solving Eqs. (11) we can determine the 6 unknowns A_1, A_2, A_3, φ_1, φ_2 and φ_3. To simplify further the numerical calculations, we take into account only the two terms A_1 and A_2 and neglect A_3, A_4, $\ldots$

Then Eqs. (12) are obtained by setting A_3 in Eqs. (11) equal to zero

References p. 118

$$A_1 (\cos \omega T - A_2 \beta_1 \omega \cos \phi_2) = S (\cos \phi_1 + 2\alpha\omega V_0 \sin \phi_1)$$

$$A_1 (\beta' \omega - \sin \omega T - A_2 \beta_1 \omega \sin \phi_2) = S(\sin \phi_1 - 2\alpha\omega V_0 \cos \phi_1)$$

$$A_1^2 \beta_1 \omega + A_2 [\cos (\phi_2 - 2\omega T) - 2\beta' \omega \sin \phi_2] = - S^2 \alpha\omega \cos (2\phi_1) \quad (12)$$

$$A_2 [\sin (\phi_2 - 2\omega T) + 2\beta' \omega \cos \phi_2] = - S^2 \alpha\omega \sin (2\phi_1)$$

where

$$\beta' = 2\beta_1 V_0 + \beta \qquad \phi_1 = \varphi_1 - \omega T \qquad \phi_2 = 2\varphi_1 - \varphi_2 \qquad (13)$$

RESULTS OF CAR-FOLLOWING EXPERIMENTS AND THE VALUE OF THE COEFFICIENTS OF THE DYNAMIC EQUATION

Street traffic (Traffic at low speeds)

In Japan the speed is limited, in general, to 40 km/h for passenger cars within cities. Hence the car-following experiment was performed for a speed range of 10 to 45 km/h. On the Gojo Boulevard in Kyoto City, a lead car (1950 Plymouth) and a following car (1956 Toyopet Master) were photographed with an 8mm cine-camera from the top of a roadside building. The latter vehicle was driven so as to keep a minimum safe car spacing. On the roadway the concrete pavement has transverse joints every 5 meters. The photographs were projected on a screen to measure car spacings and positions of the vehicles. Thus the correct car spacings and car speeds were computed after correcting for distortion of the photographs. Then assuming several values of the reaction time, T, the coefficients of Eq. (3) were derived by the method of least squares and the goodness of fit of the equation was determined for each value of T. The value of T for the case when the goodness of fit becomes greatest was adopted as the value of the reaction time and the other constants were determined simultaneously. The experimental section of the road was restricted to about 200 meters because the photographs had to be taken from the top of a building. That is the reason why the car-following experiment was repeated 22 times for speeds less than 45 km/h. Rough instructions concerning the amount and location of acceleration and deceleration were given to the driver of the lead car, whereas the driver of the following car was instructed to follow the lead car keeping a safe car spacing which he considered a minimum.

The observational results were dealt with as follows: car spacings as well as speeds of the lead car and the following car were calculated every 1/8 second, reaction time was altered by steps of 1/8 second, and the value of the coefficients of the dynamic equation and the value of the reaction time were determined so as to maximize the goodness of fit. In determining the values of the coefficients of the dynamic equation it is sufficiently accurate, judged from past experimental results[4], to assume $\alpha = -\beta_1$. Hence we have determined these coefficients β_1, β and b_0 by assuming $\alpha = -\beta_1$.

All of the 22 car-following experiments were performed by the same drivers, the drivers taking a rest after each experimental run. The coefficients were, as a matter of convenience, computed on the basis of the lumped total of the experiments. The results are given in Table I, which shows the highest goodness of fit, namely, 75% when the reaction time was chosen as 0.5 second.

TABLE I

RELATION OF REACTION TIME AND GOODNESS OF FIT

T [s]	0.125	0.250	0.375	0.500	0.625	0.750	0.875	1.000	1.125
Goodness of fit [%]	1.0	0.5	10.0	75.0	0.5	0.5	0.5	0.5	0.5

The values of the coefficients computed are as follows

$$\beta_1 = 0.00028 \text{ s}^2/\text{m}, \quad \beta = 0.585 \text{ s}, \quad b_0 = 4.131 \text{ m}, \quad T = 0.5 \text{ s}$$

Since β_1 is very small and the effect of the nonlinear condition hardly appears, it may be possible to deal with street traffic assuming $\beta_1 \simeq 0$. The frequency characteristics when $\beta_1 = 0$ are illustrated in Fig. 1.

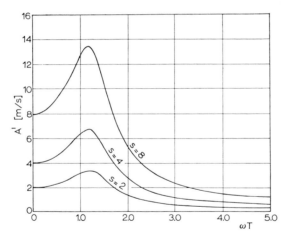

Fig. 1. Frequency characteristics for traffic with linear spacings.

Traffic on a superhighway or turnpike (Traffic at higher speed)

In Japan we have neither superhighways nor turnpikes but one is now under construction. It is expected that, when the turnpike is completed, automobile traffic will exceed by far the speed of 60 km/h which is the highest speed limit

References p. 118

on the present national highways. Thus it is necessary to investigate the change of the above mentioned coefficients of the dynamic equation for higher speeds. At this time we have no good road on which car following experiments can be performed with speeds greater than 90 km/h. So, we have practiced our car-following experiments within a speed range of 40 to 80 km/h on the test road at

Fig. 2. Test car with fifth wheel to measure speed.

Higashi-Murayama, Tokyo Prefecture. The cars used for this experiment were a 1959 Toyopet Crown as the lead car (specially equipped test car owned by the Japan Road Corporation) and a 1957 Prince Skyline as the following car. The acceleration characteristics of the following car were better than those of the lead car. Both cars had a fifth wheel to measure speed as is shown in Fig. 2. The angular speed of the fifth wheel can be recorded on recording paper. The synchronization of the recording apparatus in the lead car and in the following car was attained

TABLE II

VALUES OF COEFFICIENTS β_1 AND β

	Experiment No. 1	Experiment No. 2
β_1 [s²/m]	—0.008477	—0.003267
β [s]	0.7811	0.8270
T [s]	0.75	0.75
Goodness of fit [%]	95	25

References p. 118

by making a verbal signal through a walkie-talkie. The error of synchronization using this method was found to be less than 0.1 sec. The speeds of the two cars were observed but the distance between the cars could not be measured because of the lack of experimental apparatus. All of the coefficients, therefore, could not be determined but the values of β_1 and β were found and are given in Table II. The same vehicles were used for experiment No. 1 as for experiment No. 2 but the drivers were not the same.

Fig. 3 illustrates a portion of the speed curves obtained from the lead car and the following car in experiment No. 1. Referring to the values of β_1, β and T, we can conclude that the amplitude of the speed curve of the following car is greater than that of the lead car.

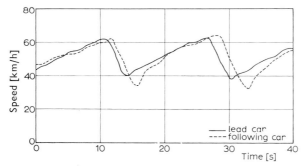

Fig. 3. Record of speed curves.

Setting $\alpha = -\beta_1, \beta_1 = p_1 T, \beta' = 2\beta_1 V_0 + \beta = pT$ in Eqs. (12) we have

$$A_1 \cos \omega T - A_1 A_2 p_1 \omega T \cos \phi_2 = S \cos \phi_1 - 2Sp_1 V_0 \omega T \sin \phi_1$$

$$A_1 p \omega T - A_1 \sin \omega T - A_1 A_2 p_1 \omega T \sin \phi_2 = S \sin \phi_1 + 2Sp_1 V_0 \omega T \cos \phi_1$$

$$A_2 \cos (\phi_2 - 2\omega T) - 2A_2 p \omega T \sin \phi_2 + A^2 p_1 \omega T = p_1 \omega T S^2 \cos (2\phi_1)$$

$$A_2 \sin (\phi_2 - 2\omega T) + 2A_2 p \omega T \cos \phi_2 = p_1 \omega T S^2 \sin (2\phi_1).$$

(14)

The frequency characteristics computed from Eqs. (14) are illustrated in Fig. 4. If the amplitude of the speed variation of the lead car is varied, the gain of the speed variation of the following car also varies according to the nonlinear characteristics of the system. But the amount of the variation is very small, as can be seen from Fig. 4, and we assume that it will be permissible to neglect it.

For both cases of the above mentioned experiments, the reaction times were observed to be very small. This fact may be understood because the driver of the following vehicle was nervously conscious of the experiment. To clarify this point we intend to perform many experiments of car following under different conditions. In all the observed results β_1 was found to be negative, but further studies are necessary to ascertain whether or not β_1 should always be negative. Since the investigations were made every 1/8 second in case (1) and every 1/4 second in case (2) in determining the reaction time, it can be considered that the optimum reaction time should exist within the time increments.

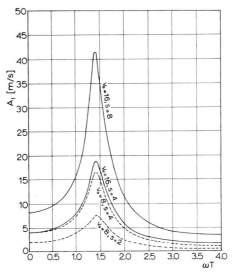

Fig. 4. Frequency characteristics for traffic with non-linear spacings.

VARIATION OF THE AVERAGE CAR SPACING

When the lead car and the following car were moving at a uniform speed and then the lead car was subjected to a sinusoidal speed variation due to a kind of disturbance, it has already been shown that the following car also undergoes a speed variation, which, however, is of a nonharmonic ocsillatory type. In this case the car spacing also varies, but does the average of the varying car space have the same value as the car space when they are moving at a uniform speed? As was studied already[3], the average car spacing under sinusoidal motion is equal to that of uniform motion so long as the system is linear.

Assuming that the speed of the lead car is given by the expression

$$V_k(t) = V_0 + S \sin \omega t \tag{15}$$

the position of the lead car is given by the equation

$$X_k(t) = V_0 t - (S/\omega) \cos \omega t + X_k(0) \tag{16}$$

where $X_k(0)$ is the initial position of the kth car. On the other hand, the speed of the following vehicle is given by Eqs. (6) and (10) as

$$V_{k+1}(t) = V_0 + \sum_{n=1}^{\infty} A_n \sin(n\omega t - \varphi_n) \tag{17}$$

and the position of the following vehicle is given by the expression

$$X_{k+1}(t) = V_0 t - \sum_{n=1}^{\infty} (A_n/n\omega) \cos(n\omega t - \varphi_n) + X_{k+1}(0) \tag{18}$$

where $X_{k+1}(0)$ is the initial position of the $(k+1)$th car. In order to determine the initial condition, we substitute Eqs. (15) to (18) into Eq. (3) and equate the constant terms on both sides, finding the relation

$$X_k(0) - X_{k+1}(0) = \alpha V_0{}^2 + \beta_1 V_0{}^2 + \beta V_0 + b_0 + \tfrac{1}{2}(\alpha S^2 + \beta_1 \sum_{n=1}^{\infty} A_n{}^2) \tag{19}$$

Furthermore the car spacing at the time t is obtained from Eqs. (16) and (18) as

$$X_k(t) - X_{k+1}(t) = \sum_{n=1}^{\infty} (A_n/n\omega) \cos(n\omega t - \varphi_n) - (S/\omega) \cos \omega t + X_k(0) - X_{k+1}(0) \tag{20}$$

Thus the average car spacing D can be expressed as

$$D = \oint \{X_k(t) - X_{k+1}(t)\} \, dt = X_k(0) - X_{k+1}(0) \tag{21}$$

The car spacing D_0 for the case where all the vehicles are moving at a uniform speed is given by

$$D_0 = \alpha V_0{}^2 + \beta_1 V_0{}^2 + \beta V_0 + b_0 \tag{22}$$

Substituting Eqs. (19) and (22) in Eq. (21) we obtain

$$D = D_0 + \tfrac{1}{2}(\alpha S^2 + \beta_1 \sum_{n=1}^{\infty} A_n{}^2) \tag{23}$$

Thus the average car spacing increases by the amount

$$\tfrac{1}{2}(\alpha S^2 + \beta_1 \sum_{n=1}^{\infty} A_n{}^2)$$

If this term has a negative value, it means that the car spacing will decrease. It is evident, according to the hitherto performed experiments, that the minimum safe car spacing in traffic moving at a uniform speed is proportional to a linear function of the speed. In this experiment too, the treatment is convenient if we set $\alpha = -\beta_1$ as was described in the preceding section. Assuming the relationship

References p. 118

$\alpha = -\beta_1$, the car spacing in traffic moving at a uniform speed becomes $D_0 = \beta V_0 + b_0$ which coincides with the experimental results obtained so far, and the average car spacing in a car-following situation with nonlinear spacings will vary by the amount

$$\tfrac{1}{2}\beta_1 \left(\sum_{n=1}^{\infty} A_n^2 - S^2 \right) \tag{24}$$

Since the quantities V_0^2 and $\tfrac{1}{2}S^2$ can be defined as the kinetic energy spectrum of the lead car and likewise $V_0^2, \tfrac{1}{2}A_1^2, \tfrac{1}{2}A_2^2, \tfrac{1}{2}A_3^2, \ldots$ can be defined as that of the following vehicle, Eq. (24) can be said to be the difference of the kinetic energy spectrum of the lead car and the following car multiplied by the coefficient β_1.

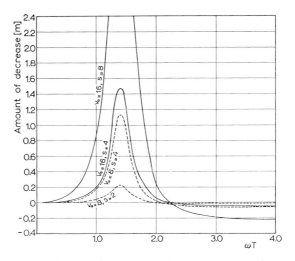

Fig. 5. Amount of decrease of the average car spacing.

To investigate the nonlinear effect for variations in the average car spacing, we computed, for one example, the variation of the average car spacing for the case $\beta_1 = -0.008477 \text{ s}^2/\text{m}$, $\beta = 0.7811 \text{ s}$, $T = 0.75 \text{ s}$. The results are given in Fig. 5.

In this example the average car spacing suffers the maximum decrease when $\omega T \cong 1.4$ as is shown in Fig. 5, and the amount of decrease becomes about 5 times greater when S is doubled. Even in the case when S has a constant value of 4 m/s, if the speed varies from 8 m/s to 16 m/s the average car spacing varies too, and the amount of variation becomes larger when V_0 is larger. On the other hand, as can be seen from Fig. 5, a frequency $f_0 = \omega_0/2\pi$ exists, such that the average car spacing is quite the same as that of uniform traffic. If $\beta_1 < 0$, the average car

References p. 118

spacing decreases for frequencies smaller than f_0 and increases for those larger than f_0. Henceforth we call this frequency f_0 the critical frequency.

In this example, since $\omega_0 T \cong 2.0$ for $V_0 = 8$ m/s and $\omega_0 T \cong 2.2$ for $V_0 = 16$ m/s, f_0 becomes 0.425 s^{-1} and 0.467 s^{-1}, respectively. As it is a very interesting phenomenon that the critical frequency varies with the speed V_0, we hope to have other opportunities to consider it.

ANGULAR VELOCITY COMPONENT OF THE SPEED VARIATION OF THE LEAD CAR

Until now we have discussed the behavior of the following car corresponding to the angular velocity ω of the speed variation of the lead car. We will now discuss the speed variation of the lead car and the kind of angular velocity component that will result if the lead car has suffered a disturbance.

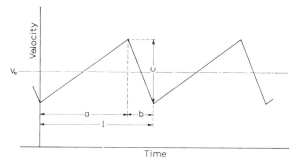

Fig. 6. Approximation of car speed.

We consider a case when the lead car has to repeat maximum acceleration and maximum deceleration alternatively because of a disturbance. In this case, the speed curve of the lead car, part of which was given already in Fig. 3, can be given as a combination of straight lines. This is illustrated schematically in Fig. 6.

If we use the coordinates shown in Fig. 6, the speed of the lead car $V_k(t)$, can be given by a Fourier series as

$$V_k(t) = V_0 + \frac{4c}{ab} \sum_{n=1}^{\infty} \left\{ \left(\frac{l}{2n\pi}\right)^2 \sin\left(\frac{2n\pi a}{l}\right) \sin\left(\frac{2n\pi t}{l}\right)\right\} \qquad (25)$$

where a and b are respectively the time necessary to accelerate or decelerate to a speed c, the period l is the sum of the constants a and b and the quantity c is the total amplitude.

The capability to accelerate or to decelerate in this case is expressed by the slope of the speed curve. The values of these quantities for the cars used in our car-following experiment were the following: $c/a = 0.44$ m/s^2, $c/b = -1.66$ m/s^2 for the lead car and $c/a = 0.89$ m/s^2, $c/b = -2.08$ m/s^2 for the following car. The minus sign means deceleration.

Since the maximum slope for the acceleration and deceleration parts of the speed curve are determined as was indicated above, the period l will be determined if an amplitude c of the speed variation is specified. Therefore, as the amplitude

TABLE III

a.   MAGNITUDE OF THE AMPLITUDE SPECTRUM FOR $c/a = 0.44$ m/s^2, $c/b = -1.66$ m/s^2

n	$c = 10$ km/h		$c = 20$ km/h		$c = 40$ km/h	
	ω	S_n	ω	S_n	ω	S_n
1	1.41	3.90	0.71	7.80	0.35	15.50
2	2.82	1.14	1.41	2.31	0.70	4.60
3	4.23	0.16	2.12	0.35	1.06	0.68
4	5.64	0.18	2.83	0.39	1.41	0.70
5	7.04	0.19	3.54	0.39	1.76	0.77
6	8.45	0.08	4.24	0.19	2.11	0.32
7	9.86	0.03	4.95	0.05	2.46	0.12
8	11.3	0.07	5.66	0.14	2.82	0.29
9	12.7	0.05	6.36	0.10	3.17	0.19
10	14.1	0.00	7.07	0.01	3.52	0.00

b.   MAGNITUDE OF THE AMPLITUDE SPECTRUM FOR $c/a = 0.89$ m/s^2, $c/b = -2.08$ m/s^2

n	$c = 10$ km/h		$c = 20$ km/h		$c = 40$ km/h	
	ω	S_n	ω	S_n	ω	S_n
1	0.79	3.75	0.40	7.48	0.20	15.10
2	1.59	1.48	0.79	2.95	0.40	5.92
3	2.38	0.62	1.19	1.25	0.59	2.46
4	3.17	0.18	1.59	0.36	0.79	0.70
5	3.96	0.04	1.98	0.08	0.99	0.18
6	4.76	0.16	2.38	0.25	1.19	0.51
7	5.55	0.12	2.78	0.25	1.39	0.50
8	6.34	0.08	3.18	0.16	1.58	0.31
9	7.14	0.03	3.57	0.05	1.78	0.08
10	7.93	0.02	3.97	0.04	1.98	0.09

of the speed variation becomes larger the angular velocity becomes smaller in general. Table III shows the magnitude of the amplitude spectrum computed for the cases $c = 10$, 20 and 40 km/h for the cars used in our experiment. We find that the primary and secondary frequencies have greater spectral densities compared to that of the other higher orders.

CONCLUSION

In this paper we have considered the characteristics of traffic flow in which the car spacing is considered as a function of the square of the speed of the lead car

References p. 118

and that of the following car. In this case, which differs from the case of linear spacing, we find the following interesting aspects about the effect of nonlinearity:

1. When a speed variation of the lead car is given as a sinusoidal wave, the motion of the following vehicle can be expressed as a superimposition of an infinite number of sinusoidal waves whose angular velocities are n times that of the lead car, where n is an integer.

2. If a traffic stream flowing at a uniform speed V_0 and with a uniform car spacing D_0 has suffered a disturbance as a consequence of the lead car being subjected to a periodic speed variation, then the average car spacing of the traffic flow will change from D_0 to D. If the speed of the lead car varies sinusoidally this change is given by Eq. (24). Thus, provided $\beta_1 > 0$, D increases when

$$\sum_{n=1}^{\infty} A_n{}^2 > S^2$$

and decreases when

$$\sum_{n=1}^{\infty} A_n{}^2 < S^2$$

Increase and decrease occur inversely if $\beta_1 < 0$. For a critical frequency f_0 which satisfies the condition

$$\sum_{n=1}^{\infty} A_n{}^2 - S^2 = 0$$

the average car spacing will remain D_0.

3. A reasonable conclusion of the preceding item (2) is that if $\beta_1 > 0$ the traffic density increases when the frequency f of the speed variation of the lead car is greater than the critical frequency f_0 and it decreases when the frequency is smaller than f_0. On the contrary, if $\beta_1 < 0$ the results are the opposite. The traffic density, therefore, begins to decrease (average car spacing increases) from the top of the queue to the rear if $\beta_1 > 0$ and $f < f_0$, or if $\beta_1 < 0$ and $f > f_0$. Thus a disturbance which looks like a shock wave propagates down the queue of vehicles.

ACKNOWLEDGMENTS

It is a great pleasure to express our thanks to Mr. S. IBUKIYAMA and others who are the members of Highway Research Section at the Civil Engineering Research Institute of the Ministry of Construction in Japan. They have kindly helped us in performing our car-following experiments.

We are indebted also to various persons of the Traffic Control Section of the Police, Kyoto Prefecture, in the preparation of the cars. Finally, we express our

References p. 118

appreciation to Mr. H. OSHIMA, assistant at Kyoto University, for his valuable
assistance in the field observation and numerical calculation.

REFERENCES

1 E. KOMETANI AND T. SASAKI, On the Stability of Traffic Flow, Report No. 1, *J. Operat. Research Japan*, 2 (1958) 11–26.
2 T. SASAKI, On the Stability of Traffic Flow, Report No. 11, *J. Operat. Research Japan*, 2 (1959) 60–79.
3 E. KOMETANI AND T. SASAKI, A Safety Index for Traffic with Linear Spacings, *Operat. Research*, 7 (1959) 704–720.
4 E. KOMETANI AND T. SASAKI, Traffic Dynamics, *Proc. Fifth Japan Road Conference*, 1959, in the press.

Appendix

Let us observe the kth and $(k + 1)$th vehicles among the vehicles moving down
a queue. As is shown in Fig. 7, the positions and speeds of the kth and $(k + 1)$th

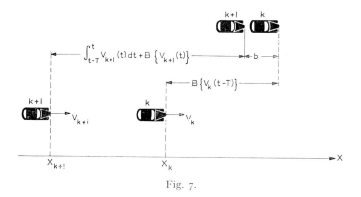

Fig. 7.

vehicles at time t are given by $X_k(t)$, $X_{k+1}(t)$, $V_k(t)$ and $V_{k+1}(t)$. If we consider
a case in which the lead car k has suddenly stopped, due to a kind of interference,
within its braking distance $B[V_k(t - T)]$, then the following vehicle $(k + 1)$ will
be able to stop after the reaction time T and thus it will move the distance given
by the expression

$$\int_{t-T}^{t} V_{k+1}(t)\,dt + B\{V_{k+1}(t)\}$$

When both vehicles have stopped it is sufficient, in order to avoid rear end
collisions, if the car spacing consists of the car length of the lead car b plus a small
amount of clearance. In order to avoid a collision of the following vehicle with a

suddenly stopped lead car, the car spacing which must be kept between cars while they are moving should not be less than

$$X_k(t-T) - X_{k+1}(t-T) = \int_{t-T}^{t} V_{k+1}(t)\mathrm{d}t + B\{V_{k+1}(t)\} - B\{V_k(t-T)\} + b \quad \text{(A.1)}$$

Since the driver of each vehicle maintains a car spacing according to his own experience, it is doubtful that the car spacing given by Eq. (A.1) is maintained by all drivers. On the right hand side of Eq. (A.1), T which is contained in the lower limit of integration of the first term, B (the factor concerning braking action) in the second and third terms and b in the fourth term, are all determined by the characteristics of the driver. So, we can assume approximately for the convenience of computation for the first term

$$\int_{t-T}^{t} V_{k+1}(t)\mathrm{d}t = \beta V_{k+1}(t) \quad \text{(A.2)}$$

and for the second, third and fourth terms

$$B\{V_{k+1}(t)\} = \beta_1 V_{k+1}^2(t)$$

$$B\{V_k(t-T)\} = \alpha_1 V_k^2(t-T)$$

and

$$b = b_0$$

Substituting Eqs. (A.2) and (A. 3) into Eq. (A.1), we have

$$X_k(t-T) - X_{k+1}(t-T) = \beta V_{k+1}(t) + \beta_1 V_{k+1}^2(t) - \alpha_1 V_k^2(t-T) + b_0 \quad \text{(A.4)}$$

If we set $-\alpha_1 \equiv \alpha$, Eq. (3) is obtained.

Single-Lane Traffic Theory and Experiment

ROBERT HERMAN AND R. B. POTTS

Research Laboratories, General Motors Corporation, Warren, Michigan
Department of Mathematics, University of Adelaide, Adelaide, South Australia

ABSTRACT

A general discussion is given of a car-following theory in which the acceleration at time t of a car attempting to follow a lead car is proportional to the relative velocity at a retarded time $t - T$ and in which the proportionality coefficient is taken to be constant, a step function or inversely proportional to the car spacing. The question of local and asymptotic stability is also discussed for traffic that obeys the above type of stimulus response equations in single lane flow with no passing allowed. A number of traffic equations of state, *i.e.* relationships between traffic flow and concentration, are described and the characteristics of the steady state flow for the model in which the proportionality coefficient depends inversely on the car spacing is compared with experiment. Follow-the-leader experiments have been carried out in various of the vehicular tunnels in New York City, in particular in the Holland and Lincoln Tunnels. These experiments yield characteristic speeds on a microscopic basis that compare very favorably with those obtained by making macroscopic measurements of overall flow and concentrations. A brief preliminary discussion is also given of the problem of measurement of macroscopic quantities such as the mean speed.

INTRODUCTION

It is useful to separate traffic problems into those concerned with city traffic, and those concerned with highway traffic. City traffic, with the complications of numerous intersections, traffic lights, turning traffic and the variety of possible traffic routes, is not readily amenable to an exact mathematical treatment but is usually described in statistical terms. Unfortunately, the statistical analysis of the vast mass of data which has been collected on city traffic has shed little light on some of the basic problems of traffic flow and stability.

In the present paper, city traffic will not be considered, but attention will be concentrated on models of highway traffic which are capable of mathematical analysis. Particular emphasis will be laid on highway traffic of the simplest kind—one-lane traffic with no passing. Such traffic conditions are quite common. Multi-lane expressways, for instance, allow dense traffic to flow with little weaving from lane to lane, so that the traffic in each lane affords an example of the simple traffic situation in mind. Of course, in special circumstances, no passing is the law. In particular, traffic through tunnels, across bridges, or on automatic highways, affords an excellent test case for the theory to be expounded in this paper.

It is important to realize that the assumption of single-lane no-passing traffic has wider and wider application as the highway system in a country improves. It is interesting to illustrate this by comparing the congestion of traffic on a poor highway with the simplicity of traffic on a modern expressway. In South Australia, for example, there are no limited-access highways, and when lanes are drawn on the road they are only for guidance. It is in fact illegal to pass on the 'wrong' side (the authors are unable to agree whether this is the left or the right side). Even on one of the best six-lane highways, traffic weaving is very common, the accident rate is high, and the speed limit is a mere 35 mph (strictly enforced). The traffic on this highway is very much more complex than for the John C. Lodge Expressway in Detroit for example, on which the speed limits are 40 mph minimum and 55 mph maximum. In fact, the average speed of cars on this expressway as reported by MALO, MIKA AND WALBRIDGE[1] is about 50 mph with about 10% of the cars exceeding the speed limit. The simplicity and homogeneity of the traffic flow on the modern highways in the United States often make a great impression on the visitor from overseas, and it is particularly reassuring to the traffic dynamicist, working with the simple model of a single-lane traffic highway, to see traffic of this kind without the complications of car passing, motorbikes, horse and carts, bicycles, rickshaws, pedestrians, etc.

There have been a variety of theoretical approaches to the study of traffic on a highway, using hydrodynamic analogies, kinetic theory, shockwaves, and the theory of stochastic processes. The model of single-lane no-passing traffic can be fairly completely described and analysed in terms of the motions of individual cars in the line of traffic, the motion of each car being assumed to follow some differential-difference equation called the car-following law. Such laws were first proposed by REUSCHEL[2] and PIPES[3] and it is extensions of these which form the theoretical basis of the work in this paper.

A car-following law (for example, that which signifies that the acceleration of a car at a delayed time is directly proportional to the relative speed of the car with respect to the one ahead) is a grossly simplified description of a very complicated response to certain stimuli. The response–stimulus description would in practice be a highly complicated functional of the dynamical properties of the cars and of the psychological and physical characteristics of the drivers. The functional would have to distinguish, for instance, between the teenage driver with his arm around his girlfriend and the driver with his wife in the back seat. It is perhaps surprising that such a functional, which one could expect to be of the most complicated mathematical structure, can be so well approximated by a simple continuous differential-difference equation. The reason for this is, of course, that in his desire to avoid accidents and get to his destination quickly, the driver suppresses his individuality and cooperates with others drivers by driving with a simple average type of response to a very limited number of stimuli.

Two of the most important traffic problems of today are those concerned with the desire to curb the accident rate and the desire to maintain an increase in traffic flow despite an increase in traffic concentration. The car-following theory of traffic sheds light on both of these problems. Rear-end and chain collisions (a common type of accident on highways) are discussed in terms of the local and asymptotic instability of a line of traffic to a fluctuation in motion of one car. Traffic flow, and its relation to traffic concentration, is investigated by deriving traffic equations of state from the car-following laws. These are relations between the flow and concentration, with parameters which give the optimum speed when the flow on the highway is a maximum.

Before proceeding to the details of recent research on car-following theory it is worth emphasizing the advantages in choosing a simple model which is capable of a precise mathematical analysis and which can be tested by experiment. Traffic data is often of little interest by itself and often reveals nothing which is not already obvious. But with a simple model and its exact behavior as a guide, predictions can be made, traffic experiments can be designed to test them, and the whole traffic problem becomes pregnant with meaning. The traffic dynamicist, with this detailed knowledge of a simple model as a background, becomes 'overnight' a traffic expert. This assumption can be carried to extremes, of course; a newspaper reporter once asked one of the authors whether he would advocate, on the basis of traffic theory, a compulsory blood test for alcoholic drivers!

TERMINOLOGY AND NOTATION

Flow q (*e.g.* in cars/hour).
The number of cars passing a given point on the highway in unit time.

Concentration k (*e.g.* in cars/mile).
The number of cars per unit length of highway at any instant.

Headway h (*e.g.* in seconds).
The difference in times at which successive cars pass a given point on the highway.

Spacing p (*e.g.* in feet).
The distance (front bumper to front bumper, or center to center) between successive cars at any instant.

References p. 146

Jam Concentration k_j (e.g. in cars/mile).
The number of cars per unit length of a traffic-jammed highway on which all the vehicles have been forced to stop.

Optimum Speed c (e.g. in miles/hour).
The speed when the flow is a maximum.

Time Lag T (e.g. in seconds).
Reaction time of the driver–car system.

Proportionality Coefficient α (e.g. in seconds^{-1}).
The ratio between the acceleration at a delayed time and the relative speed in the car-following law.

Characteristic Speed α_0 (e.g. in miles/hour).
Parameter defining the sensitivity for the reciprocal-spacing car-following law.

Other symbols used are:

$x_n(t)$ The position of the nth car at time t.
$\dot{x}_n(t)$ The speed of the nth car at time t.
$\ddot{x}_n(t)$ The acceleration of the nth car at time t.
L The effective length of a car.
U The common speed of the first and last cars of a line.
H The total headway of a line of cars.
P The total spacing of a line of cars.
C A dimensionless parameter defined by $C = \alpha T$.

CAR-FOLLOWING THEORY

Car-following laws

The basic assumption of the car-following theory of traffic flow along a single lane of a highway is that each vehicle of the line of traffic follows the one in front of it according to some stimulus–response law. Various such car-following laws have been proposed by PIPES[3], CHANDLER, HERMAN AND MONTROLL[4], and GAZIS, HERMAN AND POTTS[5]. Typical of these laws is that expressed by the equation

$$\ddot{x}_{n+1}(t + T) = \alpha \left[\dot{x}_n(t) - \dot{x}_{n+1}(t) \right], \qquad n = 1, 2, 3 \ldots \qquad (1)$$

In this equation, $\dot{x}_n(t)$ and $\ddot{x}_n(t)$ are the speed and acceleration of the nth car in the line at time t; T is the time lag of the driver–car system, and α a proportionality coefficient with dimensions of time^{-1}.

In this paper, three different laws will be considered. These are obtained from

Eq. (1) by choosing for the proportionality coefficient α the following functions of the spacing $p = x_n - x_{n+1}$ between two successive cars:

(i) constant α

$$\alpha(p) = \alpha = \text{constant} \tag{2}$$

(ii) step function α

$$\alpha(p) = \begin{cases} \alpha \text{ if } 0 < p < p_1 \\ \beta \text{ if } p > p_1 \end{cases} \tag{3}$$

where α, β, and p_1 are constants.

(iii) Reciprocal spacing

$$\alpha(p) = \alpha_0 / p \tag{4}$$

where α_0 is a constant, which will be called the characteristic speed for reasons that will become apparent from the discussion on the traffic equations of state.

With constant α, the car-following law (1) means that the acceleration of the following car, at a delayed time $t + T$, is proportional to the relative speed of the two cars. This law is, of course, an idealization of the complicated relation that really exists between the motions of two cars on a highway, but CHANDLER, HERMAN AND MONTROLL[4] have shown by a car-following experiment that the law is a good approximation. This simple law is also of great importance because of its amenability to mathematical analysis.

One of the undesirable features of the law using constant α is that it is independent of the car spacing. The two other choices of the α function attempt to remedy this defect. The step function is more realistic because it allows for two different values of the proportionality coefficient, depending on whether the two cars are close together (within a spacing p_1) or further apart. This could describe, in an extreme case, a driver whose reaction is a panic one when close to the car in front but more subdued when further away–in this case α would be chosen, say two or three times as great as β.

The law based on reciprocal spacing α has been dicussed by GAZIS, HERMAN AND POTTS[5] in the form

$$\ddot{x}_{n+1}(t + T) = \alpha_0 \frac{\dot{x}_n(t) - \dot{x}_{n+1}(t)}{x_n(t) - x_{n+1}(t)} \tag{5}$$

or

$$\dot{x}_{n+1}(t + T) = \alpha_0 \ln \{L^{-1}[x_n(t) - x_{n+1}(t)]\} \tag{6}$$

where L is the effective length of each car. This law is of great interest because it gives a reasonable explanation of the experimentally determined relation between the flow and concentration of traffic. In view of the importance of this law,

References p. 146

Fig. 1. Photograph of car-follower showing wire and reel unit.

an extensive program of experiments has been carried out in order to test its validity.

Car-following experiments

Two cars were used in car-following experiments of the kind described by CHANDLER, HERMAN AND MONTROLL[4]. To the front bumper of the following car was attached a small platform on which a reel and power unit were mounted as illustrated in Fig. 1. Several hundred feet of wire were wound on the reel and the end of the wire was fastened to the back bumper of the leading car. A constant wire tension was maintained by means of a slipping friction clutch. A potentiometer geared to the reel shaft provided a signal proportional to the length of wire (the car spacing is this length plus a car length). A dc generator tachometer operating from the other end of the shaft indicated reel motion from which the relative speeds can be determined. The speed and acceleration of the following car were measured independently and the total information:

car spacing $x_1(t) - x_2(t)$
relative velocity $\dot{x}_1(t) - \dot{x}_2(t)$
velocity of car-follower $\dot{x}_2(t)$
acceleration of car-follower $\ddot{x}_2(t)$

was recorded simultaneously by a recording oscillograph in the trunk of the follow-ing car (see Fig. 2). Fig. 3 is a reproduction of part of this record.

Four experiments on different roads have been made. In the first, reported by CHANDLER, HERMAN AND MONTROLL[4], eight different drivers were used in the instrumented car and asked to follow the lead car as it was driven on the test track at the General Motors Technical Center. This data was analyzed on the basis of the constant α model. For each run the correlation coefficient r for the correlation between $\ddot{x}_2\,(t + T)$ and $\alpha\,[\dot{x}_1(t) - \dot{x}_2(t)]$ for different values of T was evaluated by the method of least squares. The constants α and T assigned to a given driver are those for which r is a maximum. In light of the reciprocal spacing

Fig. 2. Photograph of the recording oscillograph and instrumentation in the trunk of the following car.

model, GAZIS, HERMAN AND POTTS[5] re-examined these car-following experiments. Fig. 4, taken from the above paper, gives a plot of α measured in seconds^{-1} against $\bar{p}^{-1}$, where $\bar{p}$ is the average spacing between the cars in feet. The straight line is a least squares fit (excluding the encircled point) with a straight line through the origin. The results indicate that there is a definite trend which would favor the

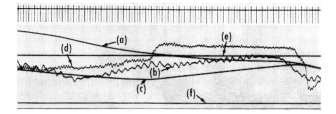

Fig. 3. Reproduction of oscillograph recording for the car-following experiment. Curve (a) represents the car spacing, (b) the relative speed, (c) the speed of the following car and (d) the acceleration of the following car. The straight line (e) is the zero reference for the relative speed and acceleration of the following car and (f) the reference for the car spacing and speed of the following car.

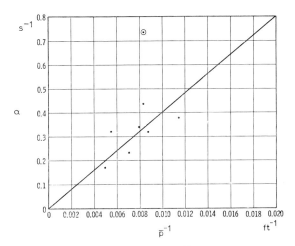

Fig. 4. Plot of proportionality coefficient α against $\bar{p}^{-1}$, the reciprocal of the average car spacing, for the car-following experiment on the General Motors Test Track. The straight line is a least-squares fit through the origin, excluding the encircled point. The gradient of the line is $\alpha_0 = 27.4$ mph.

reciprocal-spacing α function given by Eq. (4) rather than the constant α or the step function α. The gradient of the straight line gives an estimate of the characteristic speed α_0, in this case 27.4 mph. In order to further establish the reasonableness of assuming that the proportionality coefficient depends on the inverse spacing, three experiments were carried out in the Lincoln, Holland and Queens Mid-town tunnels in New York City with eleven different drivers. Sixteen, ten and four runs, respectively, were made with the two cars through the tunnels, operating under normal conditions which varied from very high to that of very low flows. From each continuous recorded run, which had an average duration of approximately four minutes, a sample of approximately one minute duration was analyzed to determine the values of α and the time lag T. Figs. 5, 6, and 7 give plots of α against $\bar{p}^{-1}$ for the three tunnels and Table I gives a summary of the results for the four experiments. Also included in this table are results for our additional experiments on the GM test track in which the two cars were driven at high speed and with violent maneuvering. This type of hard driving gave the high characteristic speed, $\alpha_0 = 82.6$ mph. In the above analysis α is assumed to be constant for each specific run, an assumption that is reasonable since it is based on the fact that the car spacing did not change very much during any one experiment.

TABLE I

RESULTS FROM CAR-FOLLOWING EXPERIMENTS

Locality	Number of runs	α_0 [mph]	T [s]
General Motors Test Track (1)	8	27.4	1.5
(2)	10	82.6	0.6
Lincoln Tunnel	16	20.3	1.2
Holland Tunnel	10	18.1	1.4
Queens Mid-town Tunnel	4	21.8	0.8

The significance of the results of these car-following experiments from the present point of view is that preference is given to the car-following law of Eq. (5) obtained from Eq. (1) using the reciprocal-spacing function α. An alternative and more precise test of the validity of this law was made by evaluating for each run the correlation coefficient r not only for the correlation between $\ddot{x}_2(t + T)$ and $\alpha [\dot{x}_1(t) - \dot{x}_2(t)]$ for different values of T, but also for the correlation between $\ddot{x}_2 (t + T)$ and

$$\alpha_0 \frac{[\dot{x}_1(t) - \dot{x}_2(t)]}{[x_1(t) - x_2(t)]}$$

for different values of T. Fig. 8 illustrates the results for one run through the Lincoln tunnel; during the run, the car spacing varied between 162 ft and 50 ft. The result indicates a time lag T of 1.6 s, corresponding to the maximum correlation, and that the correlation coefficient is significantly greater for the reciprocal-spacing α function than for constant α. The value 0.97 for the maximum value of r is remarkably close to unity giving an indication of the accuracy of the car-following law. Each run, when analyzed in this way, gives an estimate of α_0; in

References p. 146

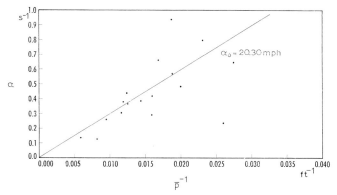

Fig. 5. Plot of proportionality coefficient α against $\bar{p}^{-1}$, the reciprocal of the average car spacing, for the car-following experiment in the Lincoln Tunnel. The straight line is a least-squares fit through the origin. The gradient of the line is $\alpha_0 = 20.3$ mph.

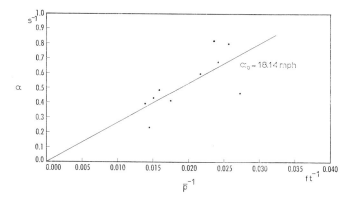

Fig. 6. Plot of proportionality coefficient α against $\bar{p}^{-1}$, the reciprocal of the average car spacing, for the car-following experiment in the Holland Tunnel. The straight line is a least-squares fit through the origin. The gradient of the line is $\alpha_0 = 18.1$ mph.

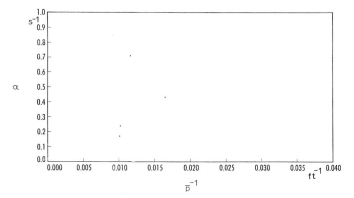

Fig. 7. Plot of proportionality coefficient α against $\bar{p}^{-1}$, the reciprocal of the average car spacing, for the car-following experiment in the Queen's Mid-town Tunnel. The limited data does not warrant a straight-line fit.

this particular case $\alpha_0 = 22.5$ mph. This analysis gives a direct test of the dependence of α on p for a given driver. For those runs where the variation of car spacing was small the data could not be used to test the dependence of α on p, but for runs with large variations in the spacing, the results were similar to those described above. This analysis thus confirms the conclusion that the car-following law with the reciprocal-spacing α function gives a very good description of the way in which cars are driven under normal traffic conditions.

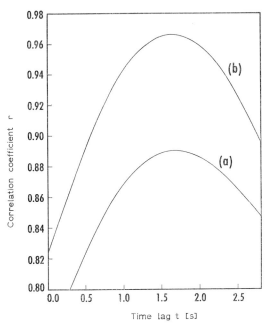

Fig. 8. Correlation coefficient, r, plotted against time lag for one run of the car-following experiment in the Lincoln Tunnel. Curve (a) corresponds to constant proportionality coefficient α and curve (b) to the reciprocal spacing coefficient $\alpha = \alpha_0/p$.

It would be expected that the response of the second car would in reality depend on whether the relative speed of the two cars is positive or negative, that is, whether the car spacing is increasing or getting smaller. An analysis of the car-following experiments showed that α_0, the characteristic speed, which is a measure of the strength of the response, is somewhat greater when the relative speed is negative, indicating a stronger response when the second car is gaining on the first car. The effect was not of such importance, however, as to justify the added complication in including this refinement to the car-following law.

It is also of interest that in certain runs in which the correlations were found to be low, a quite different stimulus–response law seemed to be obeyed. For example, for about ten seconds of one run, both cars were decelerating at the same constant rate with their speeds the same and the spacing between the cars constant. Such a situation could be described by a response giving an acceleration which is proportional to the car spacing. It is proposed to analyze such different situations in greater detail; they do suggest that in the complicated functional form of the true stimulus–response equation, different parts of the functional operate under different conditions.

Applicability of car-following laws

It is important to point out that the car-following laws can only be applied to two cars in actual traffic if the second driver is deemed to be following the first car. In actual practice some drivers, even in reasonably dense traffic, will not attempt to follow the cars preceding them and hence their cars become leaders of separate groups of cars. It is to these groups of cars, moving under the close influence of each other, that the car-following theory applies.

The criterion for deciding whether or not a car is following another is difficult to formulate precisely. The *U. S. Highway Capacity Manual*[6], for example, gives evidence of a 'free zone' in which two cars are effectively independent and an 'influenced zone' in which the cars' motions are closely dependent. The border between the zones was measured as a headway of nine seconds. From experiments carried out in Australia, GEORGE[7] has measured this headway to be six seconds. In some analyses of traffic flow, the occurrence of large gaps often causes difficulties; EDIE AND FOOTE[8] for example, exclude from their analysis gaps with headways exceeding the somewhat arbitrary figure of eleven seconds. On the basis of the car-following theory, there is no reason for choosing a criterion for car following in terms of just the headway; it could involve the spacing as well. In fact, when the car spacing is over 200 ft, the correlation coefficient calculated in the car-following experiment is rather small. It is intended to make a detailed investigation of these points in the future.

TRAFFIC STABILITY

The question of the stability of a line of traffic is an important one because of its application to accident studies. Two types of stability will be described, local and asymptotic stability. Local stability is concerned with the response of one car to a change in motion of the car in front of it and criteria can be established for this response to be oscillatory with increasing, undamped or damped amplitude, or to be non-oscillatory. Asymptotic stability, on the other hand, is concerned with the manner in which a fluctuation of the motion of the lead car is propagated down a line of traffic, and stability limits for this fluctuation to be damped as it is propagated can be determined.

Local stability

Local stability limits have been determined theoretically from the car-following law with constant proportionality coefficient α. HERMAN, MONTROLL, POTTS AND ROTHERY[9] and KOMETANI AND SASAKI[10] have shown that with this law the following situations, decribed in terms of the behavior of the spacing between the two cars, can arise:

(i)   if $C = \alpha T > \pi/2 \ (\approx 1.57)$, the spacing is oscillatory with increasing amplitude;
(ii) if $C = \pi/2$, the spacing is oscillatory with undamped amplitude;
(iii) if $1/e \ (\approx 0.368) < C < \pi/2$, the spacing is oscillatory with damped amplitude;
(iv) if $C < 1/e$, the spacing is non-oscillatory and damped.

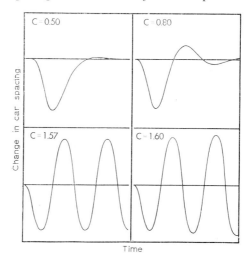

Fig. 9. Change in car spacing of two cars for different values of $C = \alpha T$ when the first car maneuvers. For $C = 0.50$ and 0.80, the spacing is oscillatory and damped, for $C = 1.57$, oscillatory and undamped, for $C = 1.60$ oscillatory with increasing amplitude.

These local stability limits were checked by a numerical solution of the car-following law for two cars when the first decelerated and then accelerated back to its original velocity. The responses for different values of C are illustrated in Fig. 9.

For the step-function α and the reciprocal-spacing α it has not been possible to determine theoretically the limits of local stability. Instead, extensive calculations have been made on the IBM 704 computer at the General Motors Research Laboratories, and from these it has been possible to gain an insight into the local stability with the exact results for the constant α case as a guide.

Figs. 10 through 14 give plots of the car spacing between two cars, originally 100 ft apart, when the first car accelerated and then decelerated (or vice versa) at the constant rate of 16.1 ft/s² for two seconds. These values of the spacing and acceleration were chosen not as representative of common traffic situations but rather as extreme cases, in order that the qualitative features of the reaction of the following car might be exaggerated. The time lag T was taken as 1 s so that

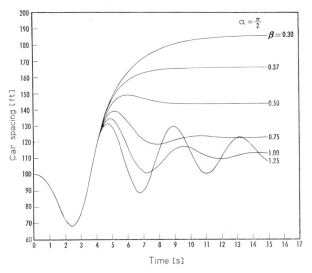

Fig. 10. Plot of car spacing *vs* time for two cars; the first decelerated for two seconds at 16.1 ft/s² and then accelerated for two seconds at 16.1 ft/s². The proportionality coefficient is the step function $\alpha = \pi/2$ for spacing less than 100 ft and various values β for a spacing greater than 100 ft.

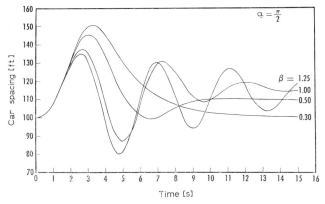

Fig. 11. Plot of car spacing *vs* time for two cars as in Fig. 10 except that the first car accelerated and then decelerated.

$C = \alpha T = \alpha$. The different curves correspond to different values for α and β for step function given by Eq. (3); the spacing p_1 separating the two different zones of response was taken as 100 ft. Figs. 10 and 11 illustrate the results for $\alpha = \pi/2$ and varying values of β. Whenever the spacing is less than 100 ft, oscillation

References p. 146

without damping occurs which throws the spacing back into the other zone, and
if β is sufficiently small, the car remains in this zone and the stability conditions
reduce to the simpler case described above. This then might adequately describe

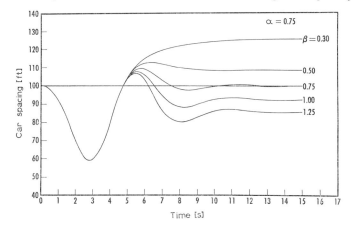

Fig. 12. Plot of car spacing *vs* time for two cars as in Fig. 10 with $\alpha = 0.75$.

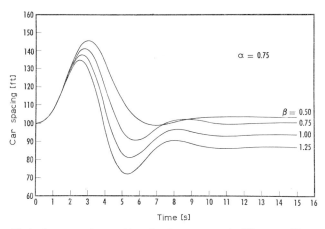

Fig. 13. Plot of car spacing *vs* time for two cars as in Fig. 11 with $\alpha = 0.75$.

a driver who responds in a much more decisive manner than usual when he is
closer than what he regards as a safe distance behind the car ahead of him, although
his reaction within this critical distance might by itself lead to local stability. In
interpreting the curve for $\alpha = \pi/2$ and $\beta = 0.30$ in Fig. 10, one can almost sense
that the driver of the following car was taken unawares by the sharp braking of
the first car and, having narrowly avoided a rear end collision, has fallen back
to a safer distance! A careful study of the other curves in Figs. 10 through 14
reveals similar local events which are of common occurrence, and suggests that

References p. 146

the step function response is similar to the way in which drivers respond to certain situations. The curves for values $\alpha < \beta$ do not seem to correspond to likely events on the road; and this is to be expected because few drivers would ever react less sensitively as they get closer to the car ahead of them.

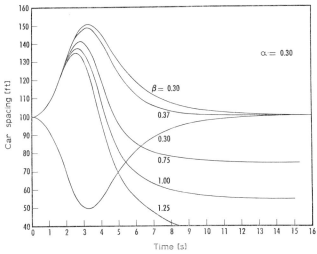

Fig. 14. Plot of car spacing *vs* time, with $\alpha = 0.3$. The five curves are for the same conditions as for Fig. 10, the other curve for the conditions as in Fig. 11.

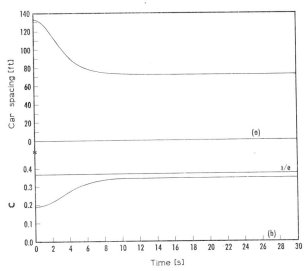

Fig. 15. (a) Car spacing *vs* time for two cars when the lead car decelerates for two seconds; the control is with the reciprocal spacing function $\alpha = \alpha_0/p$, $\alpha_0 = 17.2$ mph. The time lag $T = 1$ s. (b) Variation of $C = \alpha T$ with time; $1/e \ (\approx 0.368)$ is the local stability limit if α is constant.

For the reciprocal-spacing function $\alpha = \alpha_0/p$, it would be expected that if α remained within the limit $\alpha T < 1/e$, the motion would be locally stable. Because this type of control represents a rather safer type of control than with α constant, it would also be expected that the stability limit would be somewhat greater than $1/e$. Fig. 15 illustrates the results for two cars initially spaced at 133 ft and moving at 30 mph with $\alpha_0 = 17.2$ mph and time lag $T = 1$ s. The lead car decelerated at a constant rate for two seconds until its final speed was 19.1 mph. The figure gives a plot of the variation of the car spacing with time and the variation of $C = \alpha T = \alpha_0\,T/p$ with time. It will be noted that $C < 1/e$ throughout the

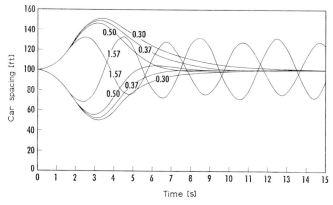

Fig. 16. Car spacing *vs* time for two cars with reciprocal spacing function $\alpha = \alpha_0/p$ for various values of α_0. The curves are labelled with the initial values of $C = \alpha T$, with $T = 1$ s. For each value of C two cases are shown, one when the first car accelerated and then decelerated and the other when the first car decelerated and then accelerated.

motion, which results in a nonoscillatory spacing. Fig. 16 illustrates the results for cars initially spaced at 100 ft, the lead car accelerating and then decelerating (or vice versa) for two seconds each at the constant rate of 16.1 ft/ s². The curves show cases of stability and instability for different values of α_0. The conclusion to be made from these results is that the characteristics of local stability in the case of α inversely proportional to spacing are not very different from the case of constant α.

Asymptotic stability

The limit for the asymptotic stability of a line of cars with respect to a fluctuation in the motion of the lead car has been investigated theoretically for the constant-α case by CHANDLER, HERMAN AND MONTROLL[4]. It was shown that if $C = \alpha T < \frac{1}{2}$, then the fluctuation is damped as it is propagated down the line, and that the rate with which the fluctuation is propagated is α^{-1} seconds per car.

Fig. 17 illustrates the results of a numerical calculation for a line of eight cars subject to a change in motion of the first car for different values of C. It is interesting to note that when $C = 1/e$ the motion is both locally and asymptotically stable. It is important to point out, however, that in the theoretical and numerical calculations of asymptotic stability, it has been tacitly assumed that the values of the proportionality coefficients for all cars in the line are the same; in practice this would not be true.

It has not been found possible to evaluate theoretically the asymptotic stability

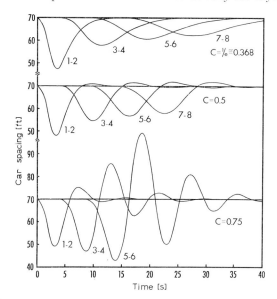

Fig. 17. Car spacings of a line of cars with constant α control for various values of $C = \alpha T$. The cars were originally spaced at 70 ft, the lead car then decelerated and then accelerated back to its original velocity, and the curves illustrate the propagation of the fluctuation down the line of cars.

limits for the step-function and reciprocal-spacing α function cases. From numerical calculations, it would seem that in the latter case, the asymptotic limit would be close to the local stability limit, because if any overshoot does take place, the value of C increases down the line of cars producing instability.

A program of experiments has been carried out to investigate the way in which a disturbance is propagated down a line of cars. Eleven cars were driven in a line down the test track at the General Motors Technical Center, and the lead car, travelling at about 40 mph, suddenly braked. The elapsed times t_6 and t_{11} between the appearance of the brake lights on the lead and the sixth and eleventh cars were measured by stop watches. Three experiments, A, B, C, were performed with five runs for the first and second and six runs for the third. In experiment A each

References p. 146

driver was instructed to react only to the brake light of the car in front of him; in B, to react to any braking stimulus; and in C, the brake lights were disconnected on all except the first and last cars. Table II gives a summary of the results for the three experiments. It is particularly interesting to note the variation in the rate at which the disturbance is propagated down the line of cars. Experiment B gave the shortest time of propagation between 0.42 and 0.58 seconds per car, while C gave the longest, about 1 second per car. These results illustrate several features. First, the difference in the time of propagation from one car to another of the disturbance with and without the brake light illustrates the importance of the brake light as a means of communicating the act of deceleration, an act which is difficult to perceive rapidly without a light. The third experiment C is suitable for a comparison with the rate of propagation as theoretically determined as this experiment corresponds closest to the simple car-following law using constant α.

TABLE II

TIMES FOR PROPAGATION OF A FLUCTUATION DOWN A LINE OF ELEVEN CARS

Run number	Experiment A		Experiment B		Experiment C
	$t_6[s]$	$t_{11}[s]$	$t_6[s]$	$t_{11}[s]$	$t_{11}[s]$
1	3.00	5.95	2.33	5.70	10.90
2	3.00	6.05	1.49	6.85	9.95
3	3.05	5.75	2.68	6.50	12.00
4	3.44	6.75	1.68	6.10	10.20
5	2.73	7.80	2.26	3.72	9.35
6	—	—	—	—	8.30
Average per car	0.61	0.65	0.42	0.58	1.01

According to the theory the propagation time is α^{-1} seconds per car so that the experiment indicates a value of 1 for α. This is just the value obtained by HELLY[11] from an analysis of a similar experiment performed by FORBES, ZAGORSKI AND DETERLINE[12].

MICRO- AND MACROSCOPIC TRAFFIC PROPERTIES

It is important to distinguish between two aspects of traffic phenomena. On the one hand, one can consider local or *microscopic* properties of the traffic, such as the effect of the motion of one car on those near it. Such properties have been described by the car-following laws discussed above. On the other hand, one can consider overall or *macroscopic* properties of the traffic, such as the relation between the flow q and concentration k of traffic. One measures the flow of traffic, that is, the number of cars passing a given point in unit time, by counting the number of cars which pass a point in a reasonably long time (say, at least a

References p. 146

minute, not just a few seconds), and the concentration, the number of cars on a unit length of road, by counting the number of vehicles on a reasonable length of road (say, at least a quarter of a mile). Measurement of q and k for short times or short lengths of roads lead to greatly fluctuating and meaningless values; in a sense q and k represent 'average' values which are meaningful from a practical point of view.

Traffic equation of state

It is useful to emphasize the macroscopic nature of the quantities q and k by referring to the somewhat analogous properties of a gas, such as pressure, volume and temperature. Pressure, for example, is computed from the average number

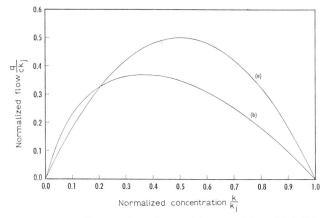

Fig. 18. Graphs of the two traffic equations of state: (a) $q = 2\,ck\,(\mathrm{1} - k/k_j)$; (b) $q = ck\,\ln\,(k_j/k)$; with $c = 17.2$ mph, $k_j = 228$ cars per mile.

of molecular collisions on the containing wall over a long (not a short) interval of time. The relation between the pressure, volume and temperature is called the gas equation of state and it is a useful analogy to call the relation between the traffic flow and concentration the traffic equation of state.

One of the fundamental problems, in traffic dynamics, is the theoretical derivation and experimental verification of this equation of state. Two equations of state which have been proposed are the following:

GREENSHIELDS' equation[13]	$q = 2\,ck\,(\mathrm{1} - k/k_j)$	(7)
GREENBERGS' equation[14]	$q = ck\,\ln\,(k_j/k)$	(8)

In each of these equations, c is the optimum speed when q is a maximum, and k_j is the concentration when the highway is jammed. These two equations are plotted for typical values of c and k_j in Fig. 18. Both equations give zero flow for concentrations $k = 0$ and $k = k_j$ and for all values of q less than the maximum, there are two possible values of k. Many experiments have been made to test these laws; typical of these are the original experiments of GREENSHIELDS[13] and the experiments of EDIE AND FOOTE[8] on the flow of traffic through the New York tunnels.

GREENSHIELDS obtained his equations by assuming that there is a linear relation between the speed of traffic and the concentration. This is only approximately true and for situations covering large changes in traffic concentrations, GREENBERG[14] has shown that his equation is a better fit to the experimental data. His derivation of the equation was based on a hydrodynamical model of traffic.

Just as the gas equation of state can be derived from the microscopic law of molecular interaction for two molecules, so it might be expected that the traffic equation of state could be derived from the 'microscopic' car-following law governing the motion of two cars. From the car-following law based on the reciprocal-spacing α function, GAZIS, HERMAN AND POTTS[5] have obtained the traffic equation of state

$$q = \alpha_0 k \ln(k_j/k) \tag{9}$$

which can be identified with GREENBERG's law (8) by putting $\alpha_0 = c$. It is thus possible to relate the macroscopic property of c, the optimum speed, measured from experiments on many cars travelling on a highway, to the characteristic speed α_0 which can be measured from the behavior of two cars driving along the same highway. To test this theory, the results obtained in the car-following experiments described in that section, have been compared in Table III with results obtained by EDIE[15] for the Holland and Lincoln tunnels.

TABLE III

CHARACTERISTIC SPEEDS

	α_0 [mph]	c [mph]
Lincoln Tunnel	20	17
Holland Tunnel	18	15
Queens Mid-town Tunnel	~22	25
GM Test Track (1)	27	
(2)	83	

When one considers the complicated nature of actual traffic and the simplicity of the car-following theory, the agreement between the two sets of results seems almost fortuitous. It is certainly astonishing that the one parameter c which is used to characterize the flow along a particular highway can be identified with the one parameter α_0 used to characterize the way in which a driver actually drives along that highway[1]. Although it has not been possible to compute an

References p. 146

accurate value of α_0 for the Queens Mid-town Tunnel, the results illustrated in Fig. 7 do suggest a value greater than that for the Lincoln tunnel. No value of c has been obtained for the General Motors Test Track, but it would be expected, of course, that the optimum speed on such an ideal highway would be greater than those for the New York tunnels.

The values of c have also been estimated from the car-following experiments by computting $\bar{u}$, the average speed of the second car, and $\bar{k}$, the average concentration (obtained from the average car spacing). The relation $\bar{u} = c \ln(k_j/\bar{k})$ is equivalent to GREENBERG's equation of state, and in Fig. 19 the values of $\bar{u}$ for

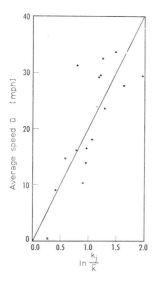

Fig. 19. Plot of average speed $\bar{u}$ against $\ln(k_j/\bar{k})$ where $\bar{k}$ is the average concentration and k_j is the jam concentration (taken as 220 cars per mile). The results are taken from the Lincoln Tunnel car-following experiment. The gradient of the least-squares line corresponds to $c = 20.1$ mph.

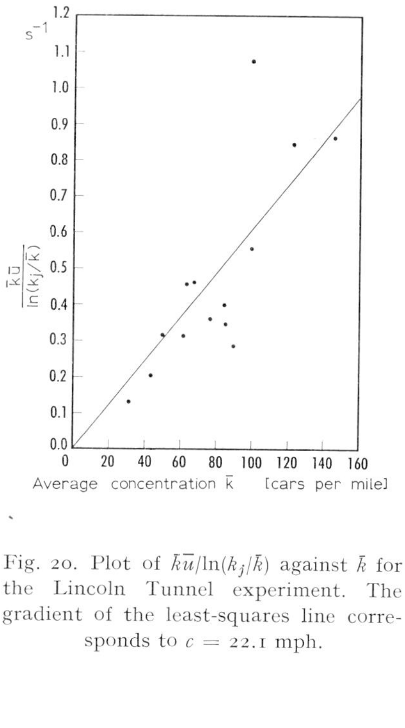

Fig. 20. Plot of $\bar{k}\bar{u}/\ln(k_j/\bar{k})$ against $\bar{k}$ for the Lincoln Tunnel experiment. The gradient of the least-squares line corresponds to $c = 22.1$ mph.

the Lincoln tunnel experiment are plotted against $\ln(k_j/\bar{k})$, with $k_j = 220$ cars per mile. The straight line, a least-squares fit going through the origin, gives a value of $c = 20.1$ mph. Fig. 20 gives a plot of $\bar{k}\bar{u}/\ln(k_j/\bar{k})$ against $\bar{k}$ and this can be compared directly with Fig. 5; the gradient of the least-squares line gives the value $c = 22.1$ mph. The general agreement between the values of c obtained from such different interpretations of the car-following law with the reciprocal-spacing α function gives further support for the conclusion that this law gives a good description of average traffic behavior. Note that in Figs. 19 and 20 a f w points coalesce. In all, sixteen experiments were included in each of the above graphs.

Measurement of macroscopic quantities

The headway h, the spacing p, and vehicle speed u are local or microscopic properties as contrasted to the macroscopic quantities q and k. Even if all vehicles are moving with constant (but differing) speeds, h and p vary with position on the highway and with time. Macroscopic space averages can be defined in terms of q (cars per hour) and k (cars per mile);

thus

$$\text{mean speed [mph]} \qquad \bar{u} = q/k \qquad\qquad (10)$$

$$\text{mean headway [s]} \qquad \bar{h} = 3600/q \qquad\qquad (11)$$

$$\text{mean spacing [ft]} \qquad \bar{p} = 5280/k \qquad\qquad (12)$$

In practice, the flow q or concentration k have often been computed from time averages of speeds, headways and spacings taken at a single point of a highway and using the above formulae, but as pointed out by WARDROP[16], time averages are not the same as space averages and can lead to incorrect results. The same criticism can be levelled at methods based on time averages of headways for given speed classes and time averages of speeds for given headway classes which are often used. It would seem necessary that a determination of the macroscopic quantities q and k for single-lane traffic should take into account that passing is not allowed. Even WARDROP's formula for the space mean speed as the harmonic mean of the speeds determined at one point of the highway assumes implicitly that the speed distribution of the cars remains constant. But this assumption would almost always be incompatible with that of no passing for the period of time over which the average speed has to be measured. Experiments to illustrate this have been performed on a highway in Adelaide, South Australia. From the speeds and spacings of cars determined at one point of a highway, trajectories have been drawn on the assumption that the cars have been moving with constant speed over the period for which the measurements were made. The many crossings of the trajectories, which could be interpreted as passings if allowed, indicate of course, that in the absence of numerous accidents some cars must alter speed considerably.

EDIE AND FOOTE[8] have pointed out the desirability of direct measurements of the flow q and the concentration k and measurements are being made in Adelaide which are based on the following theory. Consider a line of $n + 1$ cars on a single, lane no-passing highway, and suppose that the first and last vehicles move with the same constant speed U. Then both the headway H and the spacing P, between the first and last car are constant in space and time. It follows that, if U is in miles per hour and H is in seconds, the values of the flow and concentration of that particular line of vehicles are given by

$$q = 3600 \ n/H \text{ cars per hour} \tag{13}$$

$$k = 3600 \ n/(UH) \text{ cars per mile} \tag{14}$$

The spacing P in feet is given by

$$P = 5280 \ UH/3600 \text{ ft} \tag{15}$$

Provided that the first and last vehicles move with the same constant speed, the values of q and k will be the same whenever and wherever they are measured. The intermediate vehicles can accelerate or decelerate so long as they do not pass (in fact, it is actually sufficient that they do not pass the first or fall behind the last vehicle; they could pass each other!).

In terms of U, H, and P, the space averages given by Eqs. (10) (11) and (12) are

$$\bar{u} = U \tag{16}$$

$$\bar{h} = H/n \tag{17}$$

$$\bar{p} = P/n \tag{18}$$

It is interesting to note that the space mean speed $\bar{u}$ is simply U, the common speed of the first and last cars.

It might be argued that the assumption of a common constant speed for the first and last cars of the line is very restrictive and places too much emphasis on these particular cars, but at least one will be making approximate measurements with an exact theory rather than approximate measurements with an approximate theory. In practice, of course, the choice of a line of cars in which the speeds of the first and last vehicles are the same, and which is representative of the traffic state is somewhat arbitrary. It can be compared with the choice of a sufficiently large box of gas on which to experiment in order to determine the gas equation of state. From experiments with traffic, it is suggested that a line of traffic might be deemed representative of the traffic state if:

(i) the total headway between the first and last car is at least a minute;
(ii) the line of traffic is reasonably uniform, the cars travelling at much the same speed, and the gaps between successive cars not too great.

In particular, the requirement that no gap should be 'too great' causes difficulty, a difficulty faced by all who try to analyze traffic flow in terms of a theoretical model. The carfollowing theory would require that throughout the total headway time H, the cars all have to be following the car ahead. This point has been discussed in the section on the applicability of car-following laws.

Two methods have been proposed for measuring U and H in Adelaide. In one method a continuous record of the speed and headway of each vehicle is obtained as it passes a section of the highway. The record is then scanned for a line of traffic in which the first and last vehicles have the same speed and which satisfies the requirements (i) and (ii) above. This is also checked from a direct observation of the traffic from the side of the road and by control cars in the traffic. In the second method, an attempt will be made to use two controlled vehicles as the first and last cars of a line of traffic. It is proposed in this way to make a detailed investigation of the relation between the space averages, time averages and the microscopic and macroscopic properties of traffic flow.

DISCUSSION

In this paper an account of single-lane no-passing traffic has been given on the basis of a car-following theory. This theory assumes that the motion of a line of cars is governed by a car-following law relating the motion of one car to that of the car ahead. By an extensive series of car-following experiments on the test track at the General Motors Technical Center and through the tunnels in New York City, it has been shown that a relatively accurate statement of the car-following law is given by the equation

$$\ddot{x}_2(t + T) = \alpha_0 \frac{\dot{x}_1(t) - \dot{x}_2(t)}{x_1(t) - x_2(t)} \tag{19}$$

or

$$\dot{x}_2(t + T) = \alpha_0 \ln \left[L^{-1} \left(x_1 - x_2 \right) \right] \tag{20}$$

It is important not to consider this law as a dynamical one with a close physical analogy but rather to regard it as a stimulus–response equation describing the response of the driver–car system to the various stimuli. That this stimulus–response equation can be represented in such simple mathematical form is extremely fortunate; one would expect, for more complicated traffic situations, that the equation would be a most involved functional equation depending on dynamical and psychological variables, with different factors in the functional form operating at different times to describe the sorting, coding and reaction to the information gained from a vast variety of diverse stimuli. The authors believe that a detailed analysis of the whole question of the stimulus–response equation for the driver–car system could provide the psychologists with an important and interesting field of research.

In any theory of traffic one is almost compelled to relate the theory to actual traffic conditions as they are, in order to see whether some light can be shed on the tremendous problems being faced (and partly solved by slow empirical methods)

References p. 146

by the practicing highway engineer. The car-following theory in this paper has been applied in a very simple way to two very important problems, the first concerned with traffic accidents, and the second with the relief of congestion and maintenance of flow. The theory gives limits to the values of the sensitivity of the response to stimuli and the time lag of this response; beyond these limits the theory predicts instability which can lead, *e.g.* to the type of rear-end and chain collisions, which form such a high proportion of accidents on expressways. Although these limits are calculated from a simple model, they do suggest a criterion by which to distinguish between the safe and not-so-safe driver. The theory places in the latter category the driver who attempts to respond too strongly to the stimuli he receives; the reaction of the safe driver is smoother and more relaxed. These and other similar predictions of the theory seem to be borne out in practice and could with advantage, perhaps, with conclusions of other work in traffic dynamics, be included in 'better driving' manuals.

The second practical problem of the relation between the flow and density of traffic has been investigated with regard to the particular problem of the congested flow through the New York traffic tunnels. Of particular interest to those operating these tunnels is the maximum flow the tunnels can handle. The optimum speed, the speed when the flow is a maximum, has been measured in the various tunnels from the passage of many cars through the tunnels. The car-following theory predicts that this optimum speed is precisely the one parameter α_0 appearing in the car-following law (5). The values of this constant, which have been computed from the passage of the two instrumented cars through the tunnels, compare very favorably with the values obtained for the optimum speed. Thus it appears that the car-following theory gives a reasonably accurate description of some of the simple properties of tunnel flow.

In conclusion, the authors wish to stress their belief in a combined theoretical and experimental approach to the problems of traffic dynamics. Here is a field of research of undoubted practical importance which will require all the ingenuities of the mathematician, physicist, psychologist and engineer.

ACKNOWLEDGEMENTS

The authors are indebted to many of their colleagues. We wish to thank Miss P. BAKEY, Mr. R. BILDSON and Mr. R. W. ROTHERY for their aid in the numerical analysis of experimental data and to Mr. ROTHERY and Mr. R. W. REAM for collecting much of this data. We are particularly indebted to Mr. L. C. EDIE and his colleagues at the Port of New York Authority, without whose enthusiastic cooperation our experiments in the New York tunnels would have been impossible. We also thank Dr. D. C. GAZIS, Mr. R. W. ROTHERY and Mr. R. MORISON of this Laboratory, and Mr. D. H. TYLER of the Department of Civil Engineering, University of Adelaide, for the many stimulating discussions we have had on most of the topics considered in this paper.

REFERENCES

1 A. F. MALO, H. S. MIKA AND V. P. WALBRIDGE, Traffic Behavior on an Urban Expressway; presented at the *38th Annual Meeting of Highway Research Board*, January 5–9, 1959, Washington, D. C.

2 A. REUSCHEL, *Z. österr. Ingr. Architekt. Vereines*, 95 (1950) 59–62, 73–77, *Österr. Ingr. Arch.*, 4 (1950) 193–215.

3 L. A. PIPES, An Operational Analysis of Traffic Dynamics, *J. Appl. Phys.*, 24 (1953) 274–281.

4 R. E. CHANDLER, R. HERMAN AND E. W. MONTROLL, Traffic Dynamics: Studies in Car Following, *Operat. Research*, 6 (1958) 165–184.

5 D. C. GAZIS, R. HERMAN AND R. B. POTTS, Car-Following Theory of Steady-State Traffic Flow, *Operat. Research*, 7 (1959) 499–505.

6 O. K. NORMAN AND W. P. WALKER, *Highway Capacity Manual*, U. S. Department of Commerce, Bureau of Public Roads, 1950.

7 H. P. GEORGE, Methods of Measuring Traffic Volumes, Speed Delays, Traffic Capacities; paper presented at *Summer School of Traffic Eng., Univ. of Melbourne*, Australia, 1955.

8 L. C. EDIE AND R. S. FOOTE, Traffic Flow in Tunnels, *Proc. Highway Research Board, 37* (1958) 334–344.

9 R. HERMAN, E. W. MONTROLL, R. B. POTTS AND R. W. ROTHERY, Traffic Dynamics: Analysis of Stability in Car Following, *Operat. Research*, 7 (1959) 86–106.

10 E. KOMETANI AND T. SASAKI, On the Stability of Traffic Flow, *J. Operat. Research (Japan)*, 2 (1958) 11–26.

11 W. HELLY, Dynamics of Single-Lane Vehicular Traffic Flow, *Ph. D. Thesis*, Mass. Inst. Technol., 1959, reprinted as *Research Rept. no. 2, Center Operat. Research*. Mass. Inst. Technol., 1959, and in R. HERMAN (ed.), *Proc. Symposium on Traffic Flow, Detroit 1959*, Elsevier, Amsterdam, 1960, p. 207.

12 T. W. FORBES, M. J. ZAGORSKI AND W. A. DETERLINE, Measurements of Driver Reactions to Tunnel Conditions and Effects on Traffic Flow, *AIR–227–57–FR–157, Am. Inst. Research*, Pittsburgh, Pennsylvania, July 10, 1957.

13 B. D. GREENSHIELDS, A Study of Traffic Capacity, *Proc. Highway Research Board, 14* (1935) 468.

14 H. GREENBERG, An Analysis of Traffic Flow, *Operat. Research*, 7 (1959) 79–85.

15 L. C. EDIE, Private communication.

16 J. G. WARDROP, Some Theoretical Aspects of Road Traffic Research, Excerpt, Part II, *Proc. Inst. Civil Engrs., 1* (June, 1952) 325–378.

Acceleration Noise and Clustering Tendency of Vehicular Traffic

E. W. MONTROLL

IBM Technical Center, Yorktown Heights, New York

ABSTRACT

Certain statistical characteristics of vehicular traffic are considered. The acceleration noise of a vehicle on an open highway is defined as the distribution function of its acceleration. The function seems to be Gaussian for drivers who do not do much passing. The dispersion of the acceleration noise of a vehicle on a good road in the absence of traffic is of the order of $\frac{1}{2}$ ft/s². In traffic it gets multiplied by a factor of about 3. The traffic amplifying factor can be used to estimate the parameter which characters the manner in which one car follows another in traffic.

Some statistical aspects of the clustering of traffic are also discussed. From observations made in the Holland Tunnel in New York, the mean size of a cluster of vehicles is determined as a function of density for single-lane traffic. In the range 0–50 cars per mile, the mean cluster sizes grow from 1 to 2.6 cars per cluster while in the range from 50–60 it increases from 2.6 to about 4.2. By the time the density reaches 70 cars per mile the traffic has essentially jammed into one very long line.

When a car is driven on an open road in the absence of traffic the operator generally attempts, consciously or unconsciously to maintain a rather uniform velocity, but never quite succeeds. His acceleration pattern as a function of time has a random appearance. An acceleration distribution function can easily be obtained from such a pattern. This distribution is generally Gaussian. The random component of the acceleration pattern is called *acceleration noise*[1].

Runs made on a section of the General Motors test track (an almost perfect roadbed) by four operators while driving in the range of 20–60 mph yielded Gaussian acceleration noise distributions with dispersions of 0.010 g $\pm$ 0.002 g, which are about 0.32 ft/s². This dispersion increases at extreme speeds, $>$ 60 mph or $<$ 20 mph.

It is quite clear that for a given driver the acceleration noise will vary considerably as he drives on different roads or under different physiological or psychological conditions. The dispersion of the acceleration noise observed in a run in the

Holland tunnel of the New York Port Authority (with no traffic interference in the lane in which the run was made) was 0.73 ft/s². Although the roadbed of the Holland tunnel is quite good, the narrow lanes, artificial lighting, and confined conditions induce a tension in a driver which is reflected in the doubling of his acceleration noise dispersion from its perfect road value. Preliminary studies of the acceleration noise associated with runs on poorly surfaced, winding country roads indicate that dispersions of 1.5–2 ft/s² are not unusual. It would be interesting to make careful observations of this dispersion on some of the winding roads which are still prevalent in England.

Both transverse and longitudinal acceleration noise exists but no measurement of the transverse (left, right) noise has been made. It would be very large on winding roads and in the pattern of drivers who change lanes frequently while driving in heavy traffic. Both components of the noise would be large in the case of intoxicated or fatigued drivers or in situations in which the attention of the driver is shared between the road and his companions. Noise measurements have not yet been made in these situations.

Now suppose that a typical vehicle–driver combination (which we abbreviate by vd) is imbedded into a doubly infinite stream of single-lane traffic which is progressing in a stable manner (stability being indentified with the damping out of a small disturbance as one progresses backward in the line of traffic from the source of the disturbance). In particular we suppose that the effect of a disturbance practically vanishes after it has been propagated back more than k vehicles. The acceleration noise of our typical driver is composed of a superposition of his natural noise (the acceleration noise of our driver in the absence of traffic), his responce to the natural noise of his predecessor, his response to the response of his predecessor to the noise of his predecessor's predecessor, etc.

If all vehicle–driver combinations have essentially the same characteristics as our typical one of interest, the acceleration noise of our vd should be independent of his position in the line. Since our vd's total noise pattern is a reflection of his own natural noise as well as some response to his k predecessor's natural noise we can expect the dispersion of his natural noise distribution to be 'traffic broadened'. Runs were made in moderately dense traffic at 35 mph on an excellent road (Woodward Avenue, Detroit) by a vd with a natural noise dispersion of 0.32 ft/s². The traffic broadened acceleration noise dispersion was a factor of 3 greater, 0.96 ft/s², the complete distribution function being that given in Fig. 1.

Since the traffic broadening of the dispersion depends on the law through which one car follows another we should be able to use broadening data to determine the various parameters which appear in the law.

Suppose the law of following is that proposed by CHANDLER, HERMAN AND MONTROLL[2]

$$\frac{dv_n(\tau)}{d\tau} = C\{-v_n(\tau - 1) + v_{n-1}(\tau - 1)\} + \alpha_n(\tau) \tag{1}$$

where

$$C = \lambda \varDelta \tag{2}$$

λ being the sensitivity of response to a velocity difference between two successive vehicles, $\varDelta$ a time lag (usually of the order of 1.5 s), and τ the time measured in units of $\varDelta$ so that if t is the time in seconds

$$t = \tau \varDelta$$

The integer n identifies a vehicle, the $(n+1)$th being the follower of the nth, v_n the velocity of the nth vehicle and $\alpha_n(\tau)$ is the natural acceleration noise of the

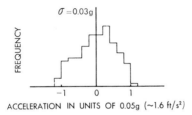

Fig. 1. Distribution of acceleration noise of a vehicle-driver combination in which the driver proceeds in traffic in a conservative manner without passing. Figure taken from HERMAN et al.[1]

nth vehicle. Suppose that the distribution of $\alpha_n(\tau)$ is the same for all vd's and that its autocorrelation function is

$$F(\eta) = E\{\alpha(\tau) \alpha (\tau + \eta)\} \equiv \lim_{T \to \infty} \frac{1}{T} \int_0^T \alpha (\tau) \alpha (\tau + \eta) \, d\tau . \tag{3}$$

The dispersion of the natural acceleration distribution function σ_0 is

$$\sigma_0{}^2 = E\{[\alpha(\tau)]^2\} \tag{4}$$

We now show how σ_0 is amplified when our vd is put into traffic.

Eq. 1 can be rewritten as

$$\dot{v}_n(\tau) + Cv_n(\tau - 1) = -Cv_{n-1}(\tau - 1) + \alpha_n(\tau) \tag{5}$$

If this equation is multiplied by a similar one obtained by replacing η by $\tau + \eta$ we find, after taking a long-time average of both sides of the resulting equation

$$E\{\dot{v}_n(\tau) \dot{v}_n(\tau + \eta)\} + CE\{\dot{v}_n(\tau) v_n (\tau + \eta - 1)\}$$
$$+ CE\{v_n (\tau - 1) \dot{v}_n (\tau + \eta)\} + C^2 E\{v_n (\tau - 1) v_n (\tau + \eta - 1)\} \tag{6}$$
$$= C^2 E\{v_{n-1} (\tau - 1) v_{n-1} (\tau - 1 + \eta)\} + E\{\alpha_n(\tau) \alpha_n (\tau + \eta)\}$$

In deriving this relation the natural acceleration noise $\alpha_n(\tau)$ of the nth vehicle was postulated to be independent of the velocity of the $(n-1)$th vehicle. Since all vd's are assumed to be identical we can disregard the subscripts on various velocities and accelerations.

Let $G(\omega)$ be the Fourier transform of the autocorrelation function of the velocity of a typical vehicle

$$G(\omega) = \frac{2}{\pi} \int_0^\infty E\{v(\tau)v(\tau+\eta)\} \cos \eta\omega \, d\eta \, . \tag{7a}$$

Then, by inversion

$$E\{v(\tau)\,v(\tau+\eta)\} = \int_0^\infty G(\omega) \cos \eta\omega \, d\omega \tag{7b}$$

from which it can easily be shown that

$$E\{v(\tau)\,v(\tau+\eta)\} = \int_0^\infty \omega\, G(\omega) \sin \eta\omega \, d\omega \tag{7c}$$

and

$$E\{\dot{v}(\tau)\,\dot{v}(\tau+\eta)\} = \int_0^\infty \omega^2\, G(\omega) \cos \eta\omega \, d\omega \tag{7d}$$

If $G_0(\omega)$ is the spectral density function of the natural velocity noise, then the autocorrelation function of the natural acceleration noise is

$$E\{\alpha(\tau)\,\alpha(\tau+\eta)\} = \int_0^\infty \omega^2\, G_0(\omega) \cos \eta\omega \, d\omega \tag{8}$$

Our problem is to compare the dispersion of the traffic broadened noise distribution, σ, which is defined by

$$\sigma^2 = E\{[\dot{v}(\tau)]^2\} = \int_0^\infty \omega^2\, G(\omega) \, d\omega \tag{9}$$

with the natural noise dispersion σ_0 defined by

$$\sigma_0^2 = E\{[\alpha(\tau)]^2\} = \int_0^\infty \omega^2\, G_0(\omega) \, d\omega \tag{10}$$

After introducing Eq. (7b), (7c), (7d), and (8) into appropriate positions in Eq. (6) we obtain (remembering that in our doubly infinite line of identical vd's the various correlation functions are independent of n)

$$\int_0^\infty [1 - 2C\omega^{-1} \sin \omega]\, \omega^2\, G(\omega) \cos \eta\omega \, d\omega = \int_0^\infty \omega^2\, G_0(\omega) \cos \eta\omega \, d\omega \tag{11}$$

Hence

$$\omega^2 G(\omega) = [1 - 2C\omega^{-1} \cos \omega]^{-1} \omega^2 G_0(\omega) \qquad (12)$$

so that from Eq. (9) the square of the dispersion of the traffic broadened noise is

$$\sigma^2 = \int_0^\infty [1 - 2C\omega^{-1} \sin s \, \omega]^{-1} \omega^2 G_0(\omega) \, d\omega \qquad (13)$$

If the spectral density $\omega^2 G_0(\omega)$ of the natural noise is peaked in the low frequency range (as generally seems to be the case) we can replace $\sin \omega$ by ω in the integrand of Eq. (13) to obtain[2]

$$\sigma \simeq (1 - 2C)^{-\frac{1}{2}} \sigma_0 \qquad (14)$$

so that the traffic broadening factor of the dispersion is $(1 - 2C)^{-\frac{1}{2}}$. An exact expression for σ/σ_0 is given in Appendix 1 when the acceleration noise is generated by a Gaussian random process.

As was shown in the previous paper of GAZIS, HERMAN AND POTTS[3], C is inversely proportional to the average spacing between vehicles, d. Hence

$$C = \alpha_0 \Delta / d \qquad (15)$$

α_0 being a constant and Δ the time lag discussed earlier. Then the spacing is

$$d = 2\alpha_0 \Delta / \{1 - (\sigma_0/\sigma)^2\} \qquad (16)$$

The validity of this formula was examined through an analysis of the data taken by HERMAN, POTTS AND ROTHERY in their car-following experiment in the Holland tunnel. The constants α_0 and Δ were found in the usual manner[2,3] and the ratio σ/σ_0 was observed. When these numbers were substituted into Eq. (16) values of d were computed which generally deviated by no more than 10% from the measured mean spacings. In an average over 9 runs the value of σ/σ_0 was 1.65.

Our traffic broadening factor was calculated under the assumption that no passing or weaving from one traffic lane to another occurred. If a driver attempts to move faster than the stream by passing, his factor should increase because of the acceleration required in the passing operation and the extra deceleration associated with squeezing back into line and thwarted attempts to pass. Generally the faster one attempts to drive relative to the mean stream velocity the greater his acceleration noise becomes.

The acceleration distribution function plotted in Fig. 1 is that of a driver who was content to follow the traffic stream on an excellent road at about 35 mph without passing. When the same driver attempted to drive 5–10 mph faster than the stream his acceleration noise pattern broadened[1] into the form exhibited in

Fig. 2. Notice that high and low acceleration wings have developed in the distribution function.

The more one weaves and attempts to travel faster than the stream the more significant his large positive and negative accelerations become.

As one tries to increase his speed above that of the stream of a given density he eventually becomes a menace on the road. An attempt might be made to use acceleration noise as a measure of the potential danger associated with a driver under various road and density conditions.

On the other hand even the most careful driver is sometimes exposed to hazardous conditions through poor visibility, bad winding roads, ice, fog, etc. Each of these would tend to broaden the acceleration noise distribution. Indeed a high frequency of accidents on a given road or involving a given driver could probably be correlated with a high acceleration noise level.

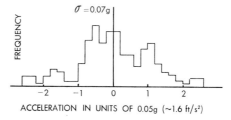

Fig. 2. Acceleration noise distribution of a driver who tries to drive five to ten miles an hour in traffic. Figure taken from HERMAN et al.[1]

Another factor which would tend to increase acceleration noise and hazards of driving is acceleration mismatch of various vehicles in a line of traffic. For example small and large cars respond in different ways to certain emergency conditions as do passenger cars and trucks. Many rear end collisions occur because of this mismatch. In fact the frequency of such collisions is greatest in a country such as Japan where such a mixed bag of traffic as bicycles, rickshaws, small and large cars and trucks are to be found on the same streets.

CLUSTERING TENDENCY

A common property of physical assemblies of many interacting elements is the existence of phase transitions. For example, as one increases the pressure of a gas at a fixed density, a pressure is suddenly achieved at which condensation occurs. At this and higher pressures the resulting condensed phase can be described as a state of a small number of very large clusters of molecules (droplets). At lower pressures the gas is composed of an enormous number of small clusters of only a few molecules each.

With this phenomenon in mind the author has searched for similar effects in

References p. 155

vehicular traffic. Their existance would imply that there should be a critical traffic density below which clusters would be very small and above which traffic moves in essentially a single dense line with very few gaps.

HERMAN, POTTS, AND ROTHERY have accumulated considerable headway and density data on traffic in the Holland tunnel of the New York Port Authority. Although this was collected for an investigation of the relation between flow rates and density, a considerable amount of it (in the density range 15–65 cars per mile) is very useful for the determination of mean cluster size as a function of density.

The definition of a cluster is somewhat arbitrary. Qualitatively a line of vehicles forms a cluster when each responds strongly to accelerations and decelerations of its predecessor. The mean spacing between vehicles in a cluster then depends on the mean stream velocity. In order to obtain a quantitative perspective of the clustering phenomenon from the data mentioned above we define a cluster to be *a line of vehicles no one of which is separated by more than 100 ft from its predecessor*. A vehicle that is more than 100 ft behind its predecessor is postulated to be the leader of a new cluster.

Although the figure of 100 ft is reasonable for the investigation of clustering tendencies at the low speeds found in the Holland tunnel, it should probably be increased in an analysis of the high-speed traffic on superhighways and freeways. Since 'safe distances' between vehicles are generally chosen on the basis of the amount of time available for a driver who wishes to accomodate to the fluctuations in speed of his predecessor, a cluster might better be defined in terms of time rather than space headways. In the material to be discussed below the vehicular speeds are of the order of 25 mph (37 ft/s) so that our 100 ft space headway would be equivalent to a 2.7 s time headway. This time would correspond to much longer range space interactions at high speeds.

In a typical run of 464 cars through the Holland tunnel at a density of 55.2 cars per mile, distribution of cluster sizes was observed to be that given in Table I.

TABLE I

DENSITY 55.2 CARS PER MILE, MEAN CLUSTER 3.01 CARS

Cluster size	1	2	3	4	5	6	7	8	9	10	11
Number	61	26	26	15	6	6	4	3	4	2	1

TABLE II

DENSITY 60.6 CARS PER MILE, MEAN CLUSTER 4.28 CARS

Cluster size	1	2	3	4	5	6	7	8	9	10	11
Number	53	40	37	16	12	4	7	6	5	2	3
Cluster size	12	13	14	17	18	19	21	24	27	28	39
Number	1	1	1	1	1	1	1	1	1	1	1

References p. 155

Even with the small change in density from 55.2 to 60.6 cars per mile it can be seen from Table II that a large number of very long lines have developed. A typical distribution of cluster sizes associated with small densities (15.1 cars per mile) is exhibited in Table III.

The results of the data given in Tables I–III, and those of several other runs are summarized in Table IV and Fig. 3. Notice the rapid variation of cluster size with density in the range 55–60 cars per mile.

TABLE III

DENSITY 15.1 CARS PER MILE, MEAN CLUSTER 1.30 CARS

Size of cluster	1	2	3	4	5	6
Number of clusters	250	49	13	5	0	1

TABLE IV

Density, cars per mile	15.1	18.3	19.3	23	50.3	55.2	60.6
Cars per cluster	1.30	1.52	1.51	1.48	2.65	3.01	4.28

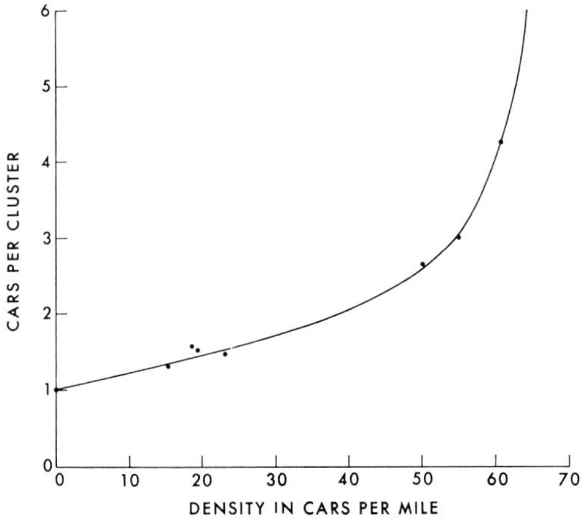

Fig. 3. Variation of average number of cars per cluster with traffic density.

In a run of 666 cars of mean density 65 cars per mile an effect similar to the existence of a two-phase region was observed. Long lines of 20–30 or more cars would appear with a density of about 70 cars per mile, then a few small clusters with lower density would develop, then another long line, etc. The range of 60–70 cars per mile seems to be analogous to the two-phase gas–liquid equilibrium which is established in molecular systems.

At densities greater than 70 cars per mile the traffic becomes essentially one long line—a gelled or jammed phase. This density corresponds to a mean front bumper to front bumper distance of 83 ft. Since fluctuations of as much as 20% from this mean value are possible without removing a car from a cluster, it is clear that the occurrence of large gaps at higher densities is rare. The gel point of some 70 cars per mile in the Holland tunnel corresponds to a peak in the flow density curve for that traffic. The mean velocity at that density is about 20 mph so that few drivers wish to travel at lower speeds and form new platoons.

It would be interesting to make cluster observations on multilane highways where passing is permitted.

ACKNOWLEDGEMENT

The author is greatly indebted to R. HERMAN, R. POTTS, and R. ROTHERY for letting him make such free use of their Holland tunnel data before its publication.

REFERENCES

1 R. HERMAN, E. W. MONTROLL, R. B. POTTS AND R. W. ROTHERY, Traffic Dynamics: Analysis of Stability in Car Following, *Operat. Research*, 7 (1959) 86.
2 R. E. CHANDLER, R. HERMAN AND E. W. MONTROLL, Traffic Dynamics: Studies in Car Following, *Operat. Research*, 6 (1958) 165.
3 D. C. GAZIS, R. HERMAN AND R. B. POTTS, Car-Following Threory of Steady-State Traffic Flow, *Operat. Research*, 7 (1959) 499.

Appendix I

TRAFFIC BROADENING OF NOISE GENERATED BY A GAUSSIAN RANDOM PROCESS

Suppose that natural acceleration noise is generated by a Gaussian random process. Then it is well known that the autocorrelation function has the form

$$E\{\alpha(t)\,\alpha(t+\eta)\} = \sigma_0{}^2\,e^{-\lambda\eta} \tag{I.1}$$

the dispersion σ_0 and the relaxation parameter λ being deducible from measurements of the variation of the acceleration with time. Then, from Eq. (8) and Eq. (I.1) the spectral-density function of the acceleration noise, $\omega^2 G_0(\omega)$, is given by

$$\omega^2 G_0(\omega) = \frac{2\sigma_0{}^2}{\pi} \int_0^\infty e^{-\lambda\eta} \cos\eta\omega \, d\eta = \frac{2\lambda\sigma_0{}^2}{\pi(\lambda^2 + \omega^2)} \tag{I.2}$$

Substituting this expression into Eq. 13 we find the traffic-broadened noise dispersion to be (since the resulting integrand is an even function of ω)

$$\sigma^2/\sigma_0{}^2 = \frac{\lambda}{\pi} \int_{-\infty}^\infty \frac{d\omega}{(\lambda^2 + \omega^2)\,(1 - 2C\,\omega^{-1}\sin\omega)} =$$
$$= (2i\lambda) \left\{ \frac{1}{2\pi i} \int_{C_0} \frac{d\omega}{(\omega + i\lambda)\,(\omega - i\lambda)\,(1 - 2C\omega^{-1}\sin\omega)} \right\} \tag{I.3}$$

where C_0 is a counterclockwise D-shaped contour over the upper half ω plane. The quantity in the bracket is the sum of the residues of the poles of the integrand. The pole at $\omega = i\lambda$ gives the contribution

$$1/(1 - 2C\lambda^{-1} \sin \lambda) \qquad (\mathrm{I}.4)$$

to σ^2/σ_0^2. As $\lambda \to 0$, as is the case when the acceleration noise is mostly composed of low frequency components, this becomes $(1 - 2C)^{-1}$ which is equivalent to Eq. (14).

An infinite set of poles are located at the zeros of

$$f(\omega) = 1 - 2C\omega^{-1} \sin \omega \qquad (\mathrm{I}.5)$$

Let ω_0 be a root of $f(\omega) = 0$. Then in the neighborhood of ω_0

$$f(\omega) \sim (\omega - \omega_0)\,\omega_0^{-1}\{1 - (4C^2 - \omega_0^2)^{\frac{1}{2}}\} \qquad (\mathrm{I}.6)$$

so that the contribution of σ^2/σ_0^2 to a pole at ω_0 is

$$\frac{2i\lambda\omega_0}{(\lambda^2 + \omega_0^2)\,[1 - (4C^2 - \omega_0^2)^{\frac{1}{2}}]} \qquad (\mathrm{I}.7)$$

As $\lambda \to 0$ these terms vanish.

A purely imaginary pole of $f(\omega)$ exists in the upper half plane when $2C < 1$. Let $\omega = iy$, then the required y satisfies

$$2Cy^{-1} \sinh y = 1 \qquad (\mathrm{I}.8)$$

If $2C \simeq 1$ as is the case in dense traffic we can approximate this root by introducing the power series for $\sinh y$ to obtain

$$1 = 2C\,(1 + y^2/6 + \ldots)$$

so that as $2C \to 1$

$$y \sim [6(1 - 2C)]^{\frac{1}{2}} \qquad (\mathrm{I}.9)$$

In order to locate the complex roots of $f(\omega)$, let $\omega = x + iy$. Then, in setting $f(\omega) = 0$ and using the fact that the real and imaginary parts of $f(\omega)$ must both vanish we obtain the two equations

$$x = 2C \sin x \cosh y \qquad (\mathrm{I}.10a)$$

$$y = 2C \cos x \sinh y \qquad (\mathrm{I}.10b)$$

Eq. (I.10a) is sketched as the solid curve in Fig. 4 and Eq. (I.10b) as the dotted curve. The intersections of the two curves occur at the zeros of Eq. (I.5). Notice that one root appears in every x interval $[2n\pi, (2n + 1)\pi]$ for $n = 1, 2, 3, \ldots$

and in every x interval $[(2n - 1)\pi, 2n\pi]$ for $n = -1, -2, -3, \ldots$ Indeed the x coordinates of the intersections are almost at the points

$$x = \pm \tfrac{1}{2}\pi \, (4n + 1) \qquad n = 1, 2, \ldots \qquad (I.11)$$

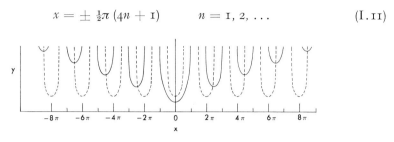

Fig. 4. Complex roots of $f(\omega) = 0$, Eqs. (I.10).

The corresponding y values are found from Eq. (I.10a) to be almost at

$$y = \cosh^{-1} \left\{ \frac{(4n + 1)}{4C} \right\} \sim \log \left\{ \frac{\pi(4n + 1)}{2C} \right\} \qquad (I.12)$$

These (x, y) values become exact in the limit as $n \to \infty$.

A Boltzmann-like Approach to the Statistical Theory of Traffic Flow

I. PRIGOGINE

The Free University, Brussels, Belgium

ABSTRACT

An integral equation for the velocity distribution of cars is studied. For suffi-

ciently small concentrations, it gives the so-called 'free speed' distribution. For

finite concentrations, the distribution is altered towards lower velocities. At some

critical concentration, the velocity distribution changes abruptly. It is suggested

that this change corresponds to going from individual to collective flow con-

The most important single function characterizing mathematically the traffic

is probably the velocity distribution function $f(v, t)$. In the limit of small concen-

trations of cars, this function depends only on the 'wishes' of the drivers, the legal

conditions as well as the nature of the highway being taken into account. It then

reduces to the 'free speed' distribution function $f^0(v)$ which has been introduced

by HAIGHT[1]. It is essentially the distribution we would have in an ideal 'dilute'

If we now consider a finite concentration each driver will be submitted to a

kind of constraint due to the interactions with the other drivers. It is clear that

this will displace the velocity distribution function towards smaller velocities.

We want to express mathematically this competition between the wishes of

the drivers and the constraints the other drivers exert on them, by an equation

somewhat similar in spirit to the fundamental Boltzmann equation of kinetic

Here also, as in the case of gases, all relevant hydrodynamical properties may

In the limit of vanishing density we assume that the time evolution of the

$$\frac{\partial f}{\partial t} = -\frac{f(v,t) - f^0(v)}{\tau} \tag{1}$$

References p. 164

which expresses that, whatever the initial distribution, after a time of order τ the free speed distribution $f^0(v)$ will be realized.

For a finite density we now write

$$\frac{\partial f}{\partial t} = -\frac{f - f^0}{\tau} + \left(\frac{\partial f}{\partial t}\right)_{coll} \tag{2}$$

Now the car 'collisions' or interactions have the characteristic property to slow down the average velocity. This can qualitatively be taken into account by assuming the following 'collision laws'

$$k + l \rightarrow 2k, \quad \text{if} \quad v_k < v_l \text{ and } \quad k + l \rightarrow 2l, \quad \text{if} \quad v_k > v_l \tag{3}$$

with a constant cross section α.

We then obtain

$$\left(\frac{\partial f}{\partial t}\right)_{coll} = C\alpha f(v) \left[\int_v^\infty f(v')\, dv' - \int_0^v f(v')\, dv'\right] \tag{4}$$

where C is the concentration.

We may then write Eq. (2) in the form

$$\frac{\partial f}{\partial t} = -\frac{f - f^0}{\tau} + C\alpha f(v) \left[1 - 2\int_0^v f(v')\, dv'\right] \tag{5}$$

We shall be mainly interested in this paper in the time-independent solution of Eq. (5).

The time-independent solution of Eq. (5) satisfies the integral equation

$$f(v) = f^0(v) + C\alpha\tau f(v) \left[1 - 2\int_0^v f(v')\, dv'\right] \tag{6}$$

Two limiting situations are of interest

$$C \rightarrow 0 \qquad\qquad f(v) = f^0(v) \tag{7}$$

$$C \rightarrow \infty \qquad 1 - 2\int_0^v f(v')\, dv' = 0 \qquad f(v) = \delta(v) \tag{8}$$

As could be expected, the velocity distribution at infinite concentration becomes independent of the 'wishes' of the driver. Then no traffic flow is at all possible.

Let us note that for $v \rightarrow 0$ Eq. (6) gives us

$$f(0) = f^0(0) + C\alpha\tau f(0) \tag{9}$$

or

$$f(0) = \frac{f^0(0)}{1 - C\alpha\tau} \tag{10}$$

which implies

$$C\alpha\tau \leq 1 \tag{11}$$

The meaning of this inequality will be discussed later in this paper.

Eq. (6) establishes a nonlinear correspondence between the probability distributions f^0 and f. In view of the problem, it may be interesting to solve this equation for one of the functions in terms of the other; Eq. (6) gives us directly f^0 in terms of f. Inversely we may obtain f in terms of f^0. Let us write Eq. (6) in the form

$$1 = \frac{f^0(v)}{f(v)} + C\alpha\tau \left[1 - 2\int_0^v f(v')\, dv'\right] \tag{12}$$

Upon differentiation we get

$$0 = (f'^0 f - f' f^0)/f^2 - 2\,C\alpha\tau f \tag{13}$$

which corresponds to the Bernouilli equation (see KAMKE[2])

$$\frac{df}{dv} = \frac{f'^0}{f^0} f - 2\frac{C\alpha\tau}{f^0} f^3 \tag{14}$$

We now write

$$f = u \exp \int_0^v \frac{f'^0}{f^0}\, dv' \tag{15}$$

to obtain the equation with separated variables, namely

$$\frac{du}{dv} = -2\frac{C\alpha\tau}{f^0} \exp\left[2\int_0^v (f'^0/f^0)\, dv'\right] . u^3 \tag{16}$$

or

$$\frac{1}{2u^2} = \frac{\gamma^2}{2} + 2\,C\alpha\tau \int_0^v \frac{dv'}{f^0} \exp\left[2\int_0^{v'} (f'^0/f)^0\, dv''\right] \tag{17}$$

where γ is an arbitrary constant. We therefore have the velocity distribution function

$$f(v) = \frac{\exp\left[\int_0^v dv'\, f'^0/f^0\right]}{\left\{\gamma^2 + 4\,C\alpha\tau \int_0^v dv' \frac{1}{f^0} \exp\left[2\int_0^{v'} (f'^0/f^0)\, dv''\right]\right\}^{\frac{1}{2}}} \tag{18}$$

The physical meaning of the constant γ is clear. When v goes to zero, Eq. (18) reduces to

$$[f(v)]_{v\to0} = \frac{1}{\gamma} \tag{19}$$

Therefore, we may also write Eq. (18) in the form

$$\frac{f(v)}{f(0)} = \frac{\exp\left[\int_0^v dv'\, f'^0/f^0\right]}{\left\{1 + 4\, C\alpha\tau f^2(0) \int_0^v dv'\, \frac{1}{f^0} \exp\left[2\int_0^{v'} (f'^0/f^0)\, dv''\right]\right\}^{\frac{1}{2}}} \tag{20}$$

Let us consider two simple examples. We first take the exponential form

$$f^0(v) = 1/v_0 \exp\left[-v/v_0\right] \tag{21}$$

We then have

$$\exp\left[\int_0^v dv' f'^0/f^0\right] = \exp\left[-v/v_0\right]$$

and

$$\frac{f(v)}{f(0)} = \frac{1}{v_0} \frac{\exp\left[-v/v_0\right]}{\{1 + 4\, C\alpha\tau v_0^2 f^2(0)\, (1 - \exp\left[-v/v_0\right])\}^{\frac{1}{2}}} \tag{22}$$

Another simple example corresponds to the 'maxwellian' free speed distribution

$$f^0(v) = \frac{4\beta^3}{\pi^{\frac{1}{2}}} v^2 \exp\left[-\beta^2 v^2\right] \tag{23}$$

For this case we obtain

$$\frac{f'^0}{f^0} = 2\left(\frac{1}{v} - \beta^2 v\right)$$

and

$$2\int_0^v dv' \left(\frac{1}{v'} - \beta^2 v'\right) = 4\,(\ln v - \tfrac{1}{2}\beta^2 v^2 - \ln 0)$$

To give a meaning to this expression it is convenient to introduce a minimum velocity v_0 and integrate everywhere from v_0 to v instead of 0 to v. We then obtain after a few elementary calculations

$$\frac{f(v)}{\beta} = \frac{1}{\varkappa} \frac{x^2\, e^{-x^2}}{\{1 + \xi\,[\Phi(x) - x\,\Phi'(x)]\}^{\frac{1}{2}}} \tag{24}$$

with

$$x = \beta v \qquad\qquad \xi = \frac{\pi C\alpha\tau}{4\varkappa^2} \qquad\qquad \Phi(x) = \frac{2}{\pi^{\frac{1}{2}}} \int_0^x e^{-x^2}\, dx$$

In these expressions $\varkappa$ is a constant (which depends on v_0) and has to be fixed by the normalization condition.

References p. 164

Again the qualitative effect of the concentration is the same. The average velocity decreases with concentration.

Let us now go back to the inequality (11). As shown by Eq. (9) it indicates that there exists a critical concentration

$$C_{cr} = 1/\alpha\tau \tag{25}$$

above which there can be no compensation between the relaxation mechanism and the 'collision' mechanism.

Of course the exact value of this critical concentration depends on the collision law (3) and on the assumption of a constant cross section α. Other more refined laws of collision are now being studied. However our qualitative result which is at the moment of primary interest is not altered. Above some critical concentration there are too many cars transferred to low velocities to be compensated by any simple restoring mechanism. The time independent integral equation has then no more solutions.

We have then to go back to the time dependent equation which indicates that $f(v)$ for $v = 0$ will increase in time till all cars are at rest and we have then precisely the δ distribution of Eq. (8).

Now this indicates that there are two regions with a discontinuous transition (somewhat like a phase transition) between them at $C = C_{cr}$. If for example we plot the average velocity as a function of concentration we obtain a curve of the form represented in Fig. 1.

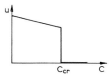

Fig. 1. Change of the average velocity
with concentration.

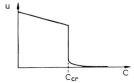

Fig. 2. Change of the average velocity
with concentration.

What is the physical meaning of these two regions? In the region below C_{cr} each car interacts successively with many cars and tries at every time to restore its 'ideal' speed. On the contrary in the region above C_{cr} the characteristic feature is that the free speed distribution does not enter at all. Now this is precisely the situation studied by CHANDLER et al.[3], HERMAN et al.[4] and GAZIS et al.[5] in their work on single-lane traffic. Here, so to speak, each car interacts always with the same one which preceeds it.

The analogy between the gaseous state for $C < C_{cr}$ and the condensed state for $C > C_{cr}$ is obvious.

A deeper distinction is however that in the first region there remains a unique correspondence between f and f^0 while in the second this is no longer so.

Therefore one could call the first region the *non-ergodic one* and the second the *ergodic one*.

It should also be noticed that the δ function in the region beyond C_{cr} is not to be taken too seriously. A more refined treatment would probably give a curve like that of Fig. 2.

More work has to be done in this connection.

An interesting feature of the 'Boltzmann' equation, Eq. (5), is that it opens the way to a 'hydrodynamics' of traffic. Indeed the natural generalization of this equation to the non-homogeneous situation is

$$\frac{\partial f}{\partial t} + v \frac{\partial f}{\partial x} = -\frac{f - f^0}{\tau} + \alpha f \left[C - 2 \int_0^v f \, dv' \right] \tag{26}$$

This equation differs from Eq. (5) by the presence of the 'flow' term $v \, \partial f / \partial x$. Also we now use the normalization

$$\int f \, dv = \int f^0 \, dv = C \, (x, t) \tag{27}$$

Here $C(x, t)$ is the local concentration at time t. Upon integration over the velocities we first derive from Eq. (26) the usual continuity equation

$$\frac{\partial C}{\partial t} + \frac{\partial}{\partial x} Cu = 0 \tag{28}$$

where u is the average local velocity.

Similarly, by multiplying Eq. (26) by the velocity and upon subsequent integration, we obtain the hydrodynamical equation

$$C \left(\frac{\partial u}{\partial t} + u \frac{\partial u}{\partial x} \right) + \frac{\partial p}{\partial x} = -\frac{C \, (u - u_0)}{\tau} + \alpha C^2 u - 2 \, \alpha v \overline{\Phi(v, x)} \tag{29}$$

Where the "pressure" p is defined by

$$p = C(\overline{v^2} - u^2) \tag{30}$$

In Eq. (29) the average velocity u_0 is calculated from the free speed distribution f^0. Also by definition

$$\overline{v\Phi(v, x)} = \int_0^\infty dv \, f(x, v) v \int_0^v dv' \, f(x, v') \tag{31}$$

The essential feature of Eq. (29) as compared to the usual equation of hydrodynamics is the existence of the 'source' term on the right hand side. We have here, of course, nothing like conservation of momentum. This is a direct consequence of the law of interaction given in Eq. (3) and is certainly a general feature of traffic hydrodynamics.

The first term on the right hand side of Eq. (29) corresponds to a kind of dynamical friction due to the difference between the ideal and the real local velocities. The physical meaning of the sum of the second and third term is similar: it corresponds also to a 'loss of momentum'.

It is clear that this non-conservation of momentum has deep reaching effects on problems like wave propagation and so on in traffic. We are at present engaged in such a study but it is too early to report here the results.

Also the appearance and disappearance of local regions of 'condensation' in the sense of the preceding section, introduces a very interesting aspect somewhat similar to the problem of nucleation in nonuniform systems.

ACKNOWLEDGEMENTS

This work originated as the consequence of many stimulating discussions the author had with Dr. R. HERMAN and Professor E. MONTROLL. The author also expresses his thanks to Dr. BALESCU and to Mr. F. ANDREWS for checking his results as well as for some calculations.

This work has been supported in part by the Research Laboratories, General Motors Corporation, Warren, Michigan.

REFERENCES

1 F. A. HAIGHT, Towards a Unified Theory of Road Traffic, *Operat. Research*, 6 (1958) 813.
2 F. KAMKE, *Differential-Gleichungen reeller Funktionen*, Göschen, Berlin, 1948, p. 26.
3 R. E. CHANDLER, R. HERMAN AND E. W. MONTROLL, Traffic Dynamics: Studies in Car Following, *Operat. Research*, 6 (1958) 165.
4 R. HERMAN, E. W. MONTROLL, R. B. POTTS AND R. W. ROTHERY, Traffic Dynamics: Analysis of Stability in Car Following, *Operat. Research*, 7 (1959) 86.
5 D. C. GAZIS, R. HERMAN AND R. B. POTTS, Car-Following Theory of Steady-State Traffic Flow, *Operat. Research*, 7 (1959) 499.

Traffic Flow Treated as a Stochastic Process

A. J. MILLER

London University, London, England

ABSTRACT

In this paper it is assumed that vehicles may be considered as travelling in clusters or queues, where a queue may be of only one vehicle, and that these queues are independent of each other in size, position and velocity. No evidence is produced to support this assumption but it is hoped to publish elsewhere some supporting evidence*. A model for traffic flow is presented in which, for mathematical convenience, queues of vehicles are represented as points. For this model a set of stochastic equations is derived assuming that the rate of catching up of queues is proportional to the product of the concentrations of those queues and their relative velocity. Overtaking is assumed to occur at a rate which is a function of queue size and velocity. The realism of the model is discussed and an adjustment is suggested to allow for the finite distances between queueing vehicles.

INTRODUCTION

Firstly, a stochastic process is a process with statistical variations and changing with respect to some suitable parameter such as time.

Now, it has long been accepted that in a traffic flow, vehicles are not randomly distributed, except perhaps in very rare circumstances when either the road is wide enough or the traffic is sufficiently light for vehicles to be able to move freely. The principal reason for nonrandomness is that because vehicles are unable to overtake freely, bunching or queueing occurs.

Let us now suppose that we have a road on which some bunching is occurring but that the traffic volume is not so high that there are no gaps. It is this kind of traffic for which this model is visualized. It will be assumed that there are no side turnings, traffic signals, intersections, pedestrians or other obstructions though the road may have bends, hills and changes in width. The idealized model proposed for the study of such traffic is as follows:

For mathematical convenience all queues in the model have zero length, and

---

* See: A. J. MILLER, A queueing model for road traffic flow, *J. Roy. Statist Soc. (B)*, 1961.

changes of speed, which will be considered as being instantaneous, occur only on either catching up another queue or upon overtaking. The word queue will generally be used instead of vehicles, a single vehicle being considered as a queue of size one. 'Splitting' of queues is not permitted and only one vehicle at a time may overtake from a queue. Overtaking is supposed to occur at a statistical rate. That is, there is a constant probability, say $\lambda \cdot \delta t$, of an overtaking in any small interval of time δt, from a given queue. This probability is independent of other overtakings but is dependent upon the size and velocity of the queue involved, the quantity of traffic flowing in the opposite direction (unless the road is a dual carriageway) and the characteristics of the road such as width, gradient and curvature. The velocity which a vehicle assumes upon overtaking will be sampled from a distribution of velocities greater than the queue velocity but so chosen that the distributions of velocities of catching up and overtaking vehicles are similar. The distribution of velocities of overtaking vehicles will be biased towards the higher velocities to allow that faster vehicles should have shorter waits on average before overtaking than the slower vehicles.

A consideration of the discrepancies between this model and reality with details of some possible adjustments, will be given after the derivation of the general stochastic equations for the process.

DEFINITIONS

Let

$\varrho_i (x, t) \delta x + o (\delta x)^2$ be the probability that at time t, there is a queue of i vehicles on the stretch of road $(x, x + \delta x)$,

$h_i (u; x, t) \delta u + o (\delta u)^2$ be the probability that the velocity of a queue of i vehicles is in the range $(u, u + \delta u)$ given that at time t the queue is at position x,

$\lambda_i (u; x, t) \delta t + o (\delta t)^2$ be the probability that a vehicle overtakes in the time interval $(t, t + \delta t)$ from a queue which is known to have i vehicles and to be at position x at time t,

$h^*(u, v; x, t) \delta u + o (\delta u)^2$ be the probability that an overtaking vehicle has a velocity in the range $(u, u + \delta u)$ given that it overtakes from a queue of velocity v which is at position x at time t.

Less formally, these functions may be considered as

ϱ_i the concentration of queues of i vehicles,

$\int_0^v h_i(u) \, . \, du$ the probability that the velocity of a queue of i vehicles is not greater than v; the velocity distributions used here are space–velocity distributions and not the time–velocity distributions usually observed in practice. (For an explanation of the difference see WARDROP[1] and MILLER[2].)

References p. 174

$\lambda_i(u)$ the overtaking rate from queues of i vehicles having velocity u,

$h^*(u, v)$ the distribution function of the velocities u of vehicles overtaking from queues of velocity v; note that this is being assumed to be independent of the queue length and the position within the queue from which the overtaking vehicle comes.

DERIVATION OF GENERAL STOCHASTIC EQUATIONS FOR THE MODEL

Equations will be found for the rate of change of the density of queues of i vehicles of velocity u, as the limit of the difference in density at times t and $t + \delta t$. For the moment, the case $i > 1$ will be considered.

During δt, the number of queues of i vehicles may increase or decrease in these ways

(i) by a vehicle overtaking from a queue of i vehicles,
(ii) by a vehicle overtaking from a queue of $i + 1$ vehicles,
(iii) by queues of $i - r$ and r vehicles catching up,
(iv) by queues of i vehicles catching up or being caught up by other queues.

The probability of occurrence for these cases is given below.

(i) At time t, the probability that there is a queue of i vehicles with velocity in the range $(u, u + \delta u)$ on the stretch of road $(x, x + \delta x)$ is

$$\varrho_i(x, t)\, \delta x\, h_i(u; x, t)\, \delta u + \text{smaller order terms.}$$

The probability that an overtaking occurs by the time $t + \delta t$ is

$$\lambda_i(u; x, t)\, \delta t + o\,(\delta t)^2$$

if there is such a queue, and so the change in probability of there being such a queue during δt due to overtaking is

$$-\lambda_i(u; x, t)\, \varrho_i(x, t)\, h;(u; x, t)\, \delta u\, \delta x\, \delta t + \text{smaller order terms.}$$

(ii) As (i) but with $i + 1$ in place of i and the opposite sign.
(iii) Let us suppose that the leading queue has $i - r$ vehicles, velocity in the range $(u, u + \delta u)$ and is in the stretch of road $(x, x + \delta x)$ at time t and that the other queue of r vehicles has velocity $u' > u$. The second queue gains a distance $(u' - u)\, \delta t$ on the first in the time δt so that it catches up if it had been within this distance at time t. The probability that such a catching up occurs then is

$$\varrho_{i-r}(x, t) \cdot h_{i-r}(u; x, t)\, \delta u\, \delta x \cdot \varrho_r(x, t)\, h_r(u'; x, t) \cdot (u' - u)\, \delta t$$

To obtain the total probability of an increase in queues of i vehicles by such catchings up, it is necessary to sum over all possible values of r and to integrate over all values of $u' > u$. This probability is then

$$\left[\sum_{r=1}^{i-1} \varrho_{i-r}(x, t) \, h_{i-r}(u; x, t) \, \varrho_r(x, t) \int_u^\infty h_r(u'; x, t) \, (u' - u) \, \mathrm{d}u'\right] \delta u \, \delta x \, \delta t$$

$$+ \text{ smaller order terms.}$$

(iv) The derivation of this probability is similar to that for (iii). The probability is

$$\left[\varrho_i(x, t) \, h_i(u; x, t) \sum_{r=1}^\infty \varrho_r(x, t) \int_0^\infty h_r(u'; x, t) \, |u' - u| \, \mathrm{d}u'\right] \delta u \, \delta x \, \delta t$$

$$+ \text{ smaller order terms.}$$

Bringing (i), (ii), (iii) and (iv) together gives the equation

$$\varrho_i(x + u\delta t, t + \delta t) \, \delta x \, . \, h_i(u; x + u\delta t, t + \delta t) \, \delta u - \varrho_i(x, t) \, \delta x \, h_i(u; x, t) \, \delta u$$

$$= \left[\sum_{r=1}^{i-1} \varrho_{i-r} \, h_{i-r}(u) \int_u^\infty \varrho_r \, h_r(u') \, (u' - u) \, \mathrm{d}u'\right] \delta u \, \delta x \, \delta t$$

$$+ \left[\lambda_{i+1} \, \varrho_{i+1} \, h_{i+1}(u) - \lambda_i \, \varrho_i \, h_i(u)\right] \delta u \, \delta x \, \delta t$$

$$- \left[\varrho_i \, h_i(u) \sum_{r=1}^\infty \int_0^\infty \varrho_r \, h_r(u') \, |u' - u| \, \mathrm{d}u'\right] \delta u \, \delta x \, \delta t$$

$$+ \text{ smaller order terms.}$$

Taking limits as $\delta t \to 0$ gives the general stochastic equations for $i > 1$

$$\frac{\mathrm{d}}{\mathrm{d}t}\left[\varrho_i \, h_i(u)\right] = \sum_{r=1}^{i-1} \varrho_{i-r} \, h_{i-r}(u) \int_u^\infty \varrho_r \, h_r(u') \, (u' - u) \, \mathrm{d}u'$$

$$\hspace{8cm} (1)$$

$$+ \lambda_{i+1} \, \varrho_{i+1} \, h_{i+1}(u) - \lambda_i \, \varrho_i \, h_i(u) - \varrho_i \, h_i(u) \sum_{r=1}^\infty \int_0^\infty \varrho_r \, h_r(u') \, |u' - u| \, \mathrm{d}u'$$

QUEUES OF ONE VEHICLE

The density of queues of one vehicle may increase or decrease in these ways:
(i) by a vehicle overtaking from a queue of two vehicles leaving a queue of one vehicle behind,
(ii) by queues of one vehicle catching up or being caught up by other queues,
(iii) by vehicles overtaking from queues when the overtaking vehicle is itself a queue of one vehicle.

As in the case of the queues of more than one vehicle, the probability of a change during the small interval of time δt will be considered. Items (i) and (ii) have been encountered already; the only new term is (iii) which is the term involving the overtaking velocity distribution $h^*(u'; u)$. Vehicles of velocity u may result from overtakings from any queues of velocity $v < u$. The probability that a single vehicle with velocity in the range $(u, u + \delta u)$ results from an overtaking within the interval of time δt from a queue of i vehicles which was within the stretch of road $(x, x + \delta x)$ at time t and had a velocity in the range $(v, v + \delta v)$, is

$$\varrho_i (x, t) \, \delta x \, h_i (v; x, t) \, \delta v \, \lambda_i (v; x, t) \, \delta t \, h^* (u; v) \, \delta u + \text{smaller order terms.}$$

Summing over all possible values of i and integrating over all values of $v < u$ gives the probability of a change in the density of queues of one vehicle due to term (iii) as

$$\left[\sum_{i=2}^{\infty} \varrho_i (x, t) \int_0^u h_i (v; x, t) \, \lambda_i (v; x, t) h^* (u; v) \, dv \right] \delta u \, \delta x \, \delta t$$
$$+ \text{smaller order terms.}$$

Bringing (i), (ii) and (iii) together gives the equation

$$\varrho_1 (x + u\delta t, t + \delta t) \, \delta x \, h_1 (u; x + u\delta t, t + \delta t) \, \delta u - \varrho_1 (x, t) \, \delta x \, h_1 (u; x, t) \, \delta u$$
$$= \lambda_2 (u; x, t) \, \varrho_2 (x, t) \, h_2 (u; x, t) \, \delta u \, \delta x \, \delta t$$
$$- \left[\varrho_1 (x, t) \, h_1 (u; x, t) \sum_{r=1}^{\infty} \int_0^\infty \varrho_r \, h_r (u') \, |u' - u| \, du' \right] \delta u \, \delta x \, \delta t$$
$$+ \left[\sum_{r=2}^{\infty} \varrho_r \int_0^u h_r (v) \, \lambda_r (v) \, h^* (u; v) \, dv \right] \delta u \, \delta x \, \delta t$$
$$+ \text{smaller order terms.}$$

Taking limits as $\delta t \to 0$ gives the general stochastic equations for $i = 1$, namely

$$\frac{d}{dt} [\varrho_1 \, h_1 (u)] = \lambda_2 \, \varrho_2 \, h_2 (u) - \varrho_1 \, h_1 (u) \sum_{r=1}^{\infty} \int_0^\infty \varrho_r \, h_r (u') \, |u' - u| \, du'$$
$$+ \sum_{r=2}^{\infty} \varrho_r \int_0^u \lambda_r \, h^* (u; v) \, h_r (v) \, dv \tag{1a}$$

REALISM OF THE MODEL IN COMPARISON WITH ACTUAL TRAFFIC

The principal departures from reality in the model appear to be these:
(i) Vehicles and queues in the model have infinitesimal length.
(ii) The model for overtaking (*i.e.* the random rate) is not a very good imitation of the overtaking behaviour in actual traffic.

(iii) Instantaneous changes of speed upon catching up or overtaking do not occur
    in real traffic.

These points will be discussed in turn and where possible corrections for them
will be suggested.

Allowance for the finite distances between queueing vehicles

In Fig. 1, the top row represents real traffic and the bottom row represents the
corresponding traffic for the model. Distances between queues are the same in
each case.

If the velocities are the same in both systems, then the rates of catching up

Fig. 1. Top and bottom rows represent real and model traffic.

will be the same and the systems are equivalent if the overtaking rates are also
the same, though the following features should be noted:

(a) The density (and also the rate of flow) of traffic in the model is higher than
    that for corresponding real traffic.
(b) If the queues in the model are randomly distributed, the corresponding queues
    in real traffic are not.
(c) Overtaking in real traffic will have to be considered as occurring when the
    overtaking vehicle has just passed the leading vehicle of the queue and not
    when the vehicle starts to overtake.

The relationship between densities in the two systems can be simply derived
as follows. Let us suppose that the average distance between the fronts of queueing
vehicles is b. Then if in the real traffic there are N vehicles in a distance X, in the
corresponding model there are N vehicles in a distance $X - Nb$. Thus if the
density in real traffic is ϱ' and it is ϱ in the model

$$\varrho = \frac{N}{X - Nb} = \frac{N/X}{1 - Nb/X} = \frac{\varrho'}{1 - \varrho'b} \qquad \varrho' = \frac{\varrho}{1 + \varrho b}$$

The overtaking model

Suppose that in real traffic there is a queue behind some fairly slow vehicle,
then the distribution of times t between consecutive overtakings is probably
something like the distribution shown in Fig. 2.

The high peak to the left is caused by those vehicles which overtake almost simultaneously with other vehicles. The broken line represents a negative exponential distribution which is approximately the distribution of intervals given by the model. To introduce a more realistic overtaking function (*e.g.* those of TANNER[3] and KOMETANI AND SASAKI[4]) into the model would require the probabilities of overtaking to be functions of the times at which vehicles caught up and the times at which other overtakings have occurred. This would make the notation and equations much more complex and would probably add little accuracy to any results or predictions from the model.

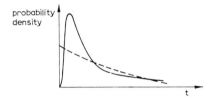

Fig. 2. Distribution of times between consecutive overtakings.

Instantaneous changes of speed

In the model, instantaneous changes of speed occur on catching up and on overtaking. To relate the model to real traffic, it has already been mentioned that overtaking must be considered as occurring when the overtaking vehicle has just passed the leading vehicle of the queue. When the overtaking vehicle passes, it has already accelerated and probably has about the speed at which it will continue and so no difficulty arises here in relating the model to real traffic. The catching up vehicle does however present a problem since unless the vehicle is able to overtake immediately, it will probably decelerate steadily to the velocity of the queue upon catching up.

In the distance–time diagram (Fig. 3), the upper continuous line represents the motion of the last vehicle of a queue and the other continuous line represents a vehicle catching up with the queue. The broken line shows the behaviour of the vehicle if it were to be obey the rules governing vehicles in the model. To transpose accurately from real traffic to the traffic of the model, the vehicle has to be considered as having caught up at time t_0 instead of t_1. In practice, if one is on the road side taking observations of the vehicles' behaviours, it is not possible to know whether or not a catching up vehicle has 'reached its t_0', and decisions as to whether vehicles are queueing have to be taken rather arbitrarily, though the following observation does provide a little guidance. If the vehicle decelerates at

References p. 174

a constant rate when catching up, then at t_0, it has cut the relative velocity between itself and the queue to exactly a half. By finding the distribution of relative velocities, the relative velocity of queue and catching-up vehicle may be used to provide a crude criterion for deciding this queueing question.

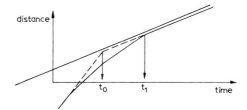

Fig. 3. Distance–time diagram.

THE FUNCTION $\lambda_n (u; x, t)$

It is hoped that eventually it will be possible to predict the overtaking rate functions λ_n for any road as a simple function (possibly a direct product) of two functions; one being a function of the road characteristics such as width, curvature, gradient, etc., and the other being a function of the quantity of traffic flowing in the opposite direction. By allowing simple theoretical conditions for overtaking, e.g., requiring a gap of at least a certain size or time interval in the opposing traffic before overtaking can occur, it is possible to construct a function of n and u to use for $\lambda_n (u; x, t)$ which is probably fairly realistic in character. The parameters in the functions can be filled in later using the results of practical experiments. Thus far, the following function has been used in computation work

$$\lambda_n (u; x, t) = a (x, t) \, \mathrm{e}^{-bu} \left(\sum_{r=1}^{n-1} \frac{1}{r^2} \right)$$

This allows for the fact that it is generally easier to overtake a slow vehicle than a fast one, and that the longer the queue is, the higher is the overtaking rate.

INDEPENDENCE OF QUEUES

Throughout the derivation of the general equations, the independence of queues, both with respect to position and velocity is implicitly assumed. It is only when events are independent that one can use the simple product law of probabilities, namely, that if p_1 and p_2 are the separate probabilities of two events occurring, then $p_1 p_2$ is the probability of both events occurring.

Let us consider one vehicle of velocity u in the traffic flow of the model. Then the queues ahead of it with velocity less than u will certainly be independent of it since that queue has always been ahead and so there has been no interaction with it. For the same reason, faster queues catching up our vehicle will be independent. It will only be the faster queues ahead and the slower ones behind which have overtaken or been overtaken by our vehicle which may not be independent and these play no part in the general equations.

But this only shows that the queues caught up by or catching up with our 'u-vehicle' are independent of it, it does not show that these queues are independent of each other. In fact, the distribution of queues in space will probably show a few fairly long queues with relatively dense concentrations of single vehicles and short queues just ahead of them. Since the model allows only one vehicle at a time to overtake, the stretch of road immediately ahead of the u-vehicle is liable to have an abnormally high proportion of queues of one vehicle just ahead of it if u is a fairly slow velocity. Thus the queues catching up our "u-vehicle" may not be quite independent of each other if there is a slow queue not far behind. The expected numbers of vehicles catching up over a small period of time will still be those predicted assuming independence and the effect of the slight interdependence of the catching up vehicles will merely be to distort slightly the distribution of the size of queue behind our 'u-vehicle'.

FURTHER WORK AND PROPOSED APPLICATION OF THIS MODEL

At present, equilibrium solutions are being obtained for sets of overtaking rates $\lambda_i(u)$. With these solutions, the values of the parameters in the λ_i will be found, by the method of maximum likelihood fitting, which give solutions most nearly approximating to data on vehicle speeds and bunching which is available. This is being done assuming that the data was obtained at points at which statistical equilibrium existed. The data which the author has available was obtained on straight flat roads well away from intersections, hills, bends, etc., and the equilibrium assumption is probably a fairly reasonable one. When the best fitting overtaking functions have been found, it is hoped to fit them, by least squares regression, as functions of the volume of opposing traffic and also, later when more data is available, as functions of the road characteristics such as width, curvature, gradient, etc. An attempt will also be made to compare predictions given by the model with conditions in actual traffic to see if any systematic deviations can be observed. A Monte Carlo experiment may be designed to discover the effect of using a more realistic overtaking model.

Dynamic solutions to the general equations will be obtained in a few special cases. It would be of interest to know how the distribution $h_1(u)$ of the velocities of single free vehicles compares with the distribution of velocities if all vehicles are free to travel at the velocities they wish to. This comparison could be found by starting from a traffic flow in which all the vehicles are travelling freely and then seeing what the eventual $h_1(u)$ is like, allowing only limited overtaking. It is also hoped to study the effect of an oscillating overtaking rate, which could correspond to a road with alternate straight and winding sections, and the case of overtaking not being possible at all.

Apart from obtaining flow–concentration curves, which will be readily obtainable from the computer results, it is hoped to make the following practical applications of the model:

(i) To study queueing behind slow vehicles to estimate the interference caused by them to traffic flow.

(ii) To study queueing at fixed-cycle traffic signals.

(iii) To study waiting time distributions for pedestrians or vehicles wishing to cross a road.

(iv) Study of the flow of traffic on a road on which there is a stretch, perhaps of bends or a hill, a tunnel, etc., on which little or no overtaking is possible.

(v) Study of the spreading of vehicles leaving traffic signals, assuming that the road ahead is clear before the lights become green.

CLOSING REMARKS

This has only been an outline of my work so far on this subject. Fuller details are given in my Manchester M. Sc. thesis and in a report which I have written. I hope also in the near future to publish a paper on these results, which I have attempted to summarize in this paper.

A considerable amount of work remains to be done including the numerical solution of equations, the testing of the goodness of fit of the model to real traffic and the application of the results. It is hoped to obtain as the result, a set of formulae and tables with which one will be able to predict desired details of traffic behaviour in any given set of circumstances.

REFERENCES

1 J. G. WARDROP, *Proc. Inst. Civil Engrs.*, 2 (1952) 325.
2 A. J. MILLER, *M. Sc. Thesis*, University of Manchester, 1959.
3 J. C. TANNER, *Dept. Sci. Ind. Research, Road Research Lab. Note* RN/2921/JCT.
4 E. KOMETANI AND T. SASAKI, *J. Operat. Research Japan*, 2 (1958) 11–26.

Experiments on Single-Lane Flow in Tunnels

L. C. EDIE AND R. S. FOOTE

The Port of New York Authority, New York

ABSTRACT

The Port of New York Authority staff for the past several years has been conducting experiments on traffic flow in their tunnels and in those of other operating agencies. Since the growth of the field of operations research, these experiments have to an increasing degree employed the more rigorous methods and approach of the physical scientist and research engineer. Statistical and deterministic models of traffic behavior have been used to guide experiments in the behavior of flow, velocity and concentration of traffic streams and to measure various parameters. Experiments are reported covering the behavior of platoons and waves in different parts of tunnels and the effect of bottlenecks on such behavior. Some experiments involve the recording on productograph tapes of vehicles' entry and exit times in measured zones for all vehicles in a stream. Other types of experiments involve stop-watch observations by observers riding in vehicles or standing on the tunnel catwalks.

INTRODUCTION

The Port of New York Authority has two tunnels under the Hudson River; namely, the Holland Tunnel comprising two tubes and the Lincoln Tunnel comprising three tubes. Each tube provides a two-lane roadway, but since lane changing is prohibited within these tunnels, each lane has the no-passing and most other features of single-lane roadways.

The vital importance of these roadways joining the center of New York City with New Jersey communities has made desirable a continuous search for methods of increasing their capacity. These two tunnels now carry close to fifty million vehicles annually and have required investments of $240 million for their construction. Rising construction costs and the increasing scarcity of relatively undeveloped land for approach roadways underscores the importance of attaining higher peak-hour lane production from existing roadways.

The search for methods of increasing lane production was first made by the operating staffs of the tunnels, who discovered improvements over the years by

References p. 192

the use of intuition and common sense alone. However, as we have found ways to employ the more advanced methods made available by mathematicians and scientists, we have done so.

The aim of this paper is to explain briefly the results of various methods of observation, experiment and analysis which have been employed by the New York Port Authority in recent years.

STATISTICAL EXPERIMENTS

The first methods employed were largely statistical. In 1952 the New York Port Authority's traffic engineering staff became interested in the high peak-hour traffic flows being attained at another tunnel in the New York area, (the Queens Midtown Tunnel.) These flows were much higher than those attained at the Lincoln Tunnel, although both tunnels had similar physical characteristics.

TABLE I

LINCOLN TUNNEL–QUEENS MIDTOWN TUNNEL PEAK HOUR FLOWS, JUNE 1952

	Lincoln (eastbound)				Queens (westbound)			
	Cars	Buses	Trucks	Total	Cars	Buses	Trucks	Total
Fast Lane	770	202	18	990	1474	3	14	1491
%	77.8	20.4	1.8	100.0	98.9	0.2	0.9	100.0
Slow Lane	809	0	152	961	1062	8	214	1284
%	84.2	0.0	15.8	100.0	82.8	0.6	16.6	100.0
Totals	1579	202	170	1951	2536	11	228	2775
%	80.8	10.3	8.9	100.0	91.4	0.4	8.2	100.0

Table I gives an analysis of the traffic carried by each tunnel in a typical weekday peak hour in 1952. The total for Lincoln Tunnel was only 1951 vehicles whereas the Queens Midtown Tunnel carried 2775 vehicles.

In an effort to understand this large difference in production, a study was made of the speed and density behavior of both tunnels under peak traffic flow. In so doing, speeds were sampled by stop-watch clocking over nine 100-ft courses located at strategic points in both tunnels. The results, which were reported by STRICK-LAND[1], were of interest with respect to differences within each tunnel in addition to differences between tunnels. When values of sampled speeds were plotted against location, speed profiles of each tunnel were obtained. Fig. 1, shows these results for the Lincoln Tunnel and Fig. 2 for the Queens Midtown Tunnel. A chart at the bottom of each figure gives the physical profile of each tunnel for ready comparison with the speed profiles. All data shown are self explanatory, except perhaps the density index, which gives mean vehicles per 100 ft.

These charts can be interpreted as revealing a tendency for bottlenecks to form at the bottom of the downgrade entrance sections where speed curves show minimal values. However, why the bottleneck did form at Lincoln, where speed dropped to less than the entrance speed, but not at Queens Midtown, where it rose high above entrance speed and then dropped only slightly, was unclear at the time.

In 1953, OLCOTT[2] became interested in comparing these tunnels using a different approach. He sampled speeds and densities at more than one location but took a continuous sample and combined all data in order to study the relationship be-

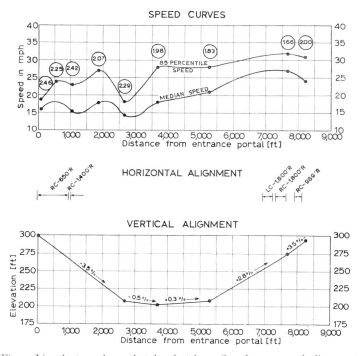

Fig. 1. Lincoln tunnel, south tube, fast lane. Speed curves and alignment.

tween speed and density at various levels of each. Using 5-minute time slices of traffic to compute mean space speeds and mean densities, he ran a linear regression analysis and found that 88% to 97% of the variability in traffic speeds were related to changes in the density of the stream. The strongest correlation coefficients were found in lanes having the highest percentage of passenger cars. Fig. 3 shows the regression line for the Lincoln Tunnel and Fig. 4 that for the Queens Midtown Tunnel. From further analysis, the various pertinent traffic flow characteristics of each tunnel listed in Table II could be deduced. This method, which results in an

References p. 192

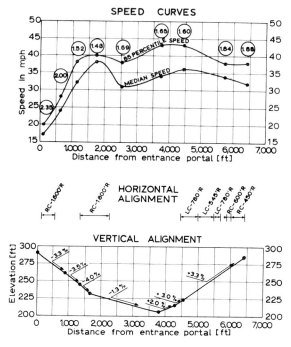

Fig. 2. Queens Midtown tunnel, north tube, fast lane. Speed curves and alignment.

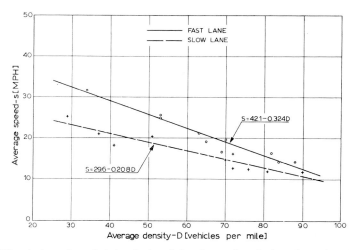

Fig. 3. Lincoln tunnel south tube, fast and slow lanes. Correlation of speed and density.

References p. 192

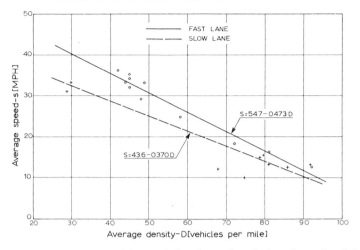

Fig. 4. Queens Midtown tunnel fast and slow lanes. Correlation of speed and density.

TABLE II

Tunnel	Lane	Range of comm. veh. [%]	Critical density [vehicles per mile]	Optimum spacing [feet]	Optimum speed [miles per hour]	Practical capacity [vehicles per hour]
Lincoln	Fast	7–17	65	81	21.0	1280–1450
Lincoln	Slow	20–35	71	75	14.8	950–1150
Queens Midtown	Fast	0	58	91	27.2	1490–1660
Queens Midtown	Slow	10–25	59	89	21.8	1100–1470

optimum speed just one-half of free speed, overestimates optimum speed and capacity, and underestimates optimum or critical density.

These early experiments yielded valuable quantitative measures and geometric pictures of flow, speed and density relationships within the two tunnels, but left largely unanswered the questions of how the relationships came about and what could be done to increase peak flow. However, during the course of these and similar observations two ideas were presented for improving flow by controls imposed on entering traffic.

The first of these had already been put into effect at the Queens Midtown Tunnel where it seemed to contribute to the superior performance observed there. At the Queens Midtown Tunnel, traffic was permitted to merge slowly and continuously from six toll lanes down to two tunnel lanes, whereas at the Lincoln Tunnel the six lanes of vehicles were alternately held and permitted to flow in rotation into the tunnel under police officer control. When the Lincoln Tunnel traffic was started and stopped, gaps frequently occurred of up to 15 s. Three gaps occurring in a minute could add up to 30 s. It was suspected that these gaps represented a direct loss in capacity. The continuous feed method was tried at the Lincoln Tunnel as one of the first experiments conducted using rigorous scientific controls.

The other experiment involved the mixing of trucks equally in both lanes rather than confining them mostly to the right-hand lanes. Analysis of the results of the two experiments was handled by non-parametric and parametric statistical methods in the absence of any flow theory.

In order to achieve significance in such statistical experiments, it was necessary to employ extensive data over periods of several months, also to investigate and correct the peak-hour flow figures for known unusual occurrences such as disabled vehicles, and finally to use other tubes and data from other years as statistical controls. Reports on the details of these studies are available[3] for anyone who might be interested in them. It was found that for the continuous traffic feed tested at the Lincoln Tunnel an increase of 73 vehicles per hour with a standard deviation of 28 was indicated, and it was found that for the mixing of trucks in both lanes tested at the Holland Tunnel a decrease of 80 with a standard deviation of 30 was indicated.

EXPERIMENTS RELATING TO THEORY

The great amount of work and other inherent difficulties in trying to cope with the tunnel flow problem by such statistical methods alone is obvious. Also these statistical methods, which have been used for many years by traffic engineers and others, have shed very little light on the dynamics of traffic streams. While much data has been collected to show the kind of performance observed for various types of roadways throughout the country, little basic knowledge has accrued therefrom. After some years of efforts to use statistical methods, we concluded that further significant improvements would depend heavily on the development of theories of traffic flow that would yield deeper and more precise understandings of the phenomena.

In seeking to develop flow theories a number of general approaches are available, including analytical, empirical and simulation methods from either the microcosmic study of single vehicles or the macrocosmic studies of average behavior. At the New York Port Authority emphasis has been placed on empirical methods of observation and controlled experiments applied to stream behavior. However, the New York Port Authority has also participated in the work by FORBES et al.[4] with 3-car experiments involving deceleration–acceleration maneuvers, by HERMAN *et al.*[5] with 2-car following experiments inside the tunnels, and work by HELLY[6] in simulating tunnel traffic on a large scale computer. The balance of this paper will be concerned with experiments conducted by the New York Port Authority staff for the dual purpose of contributing to the development of traffic flow theory, and of learning more about the particular phenomena occurring inside the tunnels so that further improvements in flow could be made.

THE FLUID FLOW MODEL

LIGHTHILL AND WHITHAM[7] suggested that an effective description of various traffic flow phenomena could be derived from assumptions of continuity and fluid behavior. A phenomenon which concerned them particularly was that of the behavior of the stream on each side of a bottleneck. Because of the previous

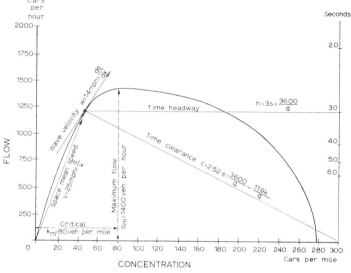

Fig. 5. Dimensional analysis of flow as a function of traffic concentration.

discovery of bottleneck sections within tunnels, the Lighthill and Whitham approach appeared most suitable as a model for further empirical studies. As suggested by these authors, the study of flow–concentration relationships at a point and wave phenomena has proven quite useful in extending knowledge of tunnel traffic flow and traffic in general.

To review the kinematic behavior of bottlenecks reference is made to Fig. 5. Here a humpback curve is shown for the average relationship between flow and concentration on a roadway with a maximum mean flow rate of 1400 vehicles per hour. For the particular curve hypothesized, a mean flow of 1200 vehicles per hour would occur at a speed of 25 mph and a concentration of 48 vehicles per mile. Furthermore, this flow state would be propagated forward in space at a speed of 14 mph, as given by the slope of the tangent to the curve. When there are two roadways in series, with flow–concentration curves of different maxima, the one with lower capacity becomes a bottleneck for the upper one, as illustrated in Fig. 6. Assuming fluid behavior, LIGHTHILL AND WHITHAM predicted that traffic upstream from the bottleneck would jump from state A to state C passing rapidly through D, thereby resulting in congested crawling traffic conditions; and downstream from the bottleneck, traffic flow conditions would move from state B to state A.

Experiments were conducted with a Simplex Productograph time clock to test

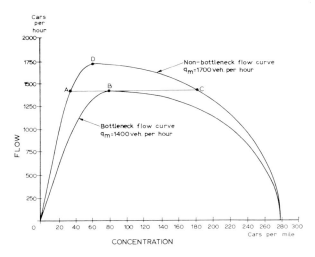

Fig. 6. Kinematic behavior of traffic bottlenecks.

these predictions. The machine, which was set up along a roadway zone of 33 ft, was actuated by two observers who depressed one of several push buttons whenever a vehicle entered or left the test zone. This classified the vehicles by type depending on which button was depressed and recorded headway and transit times within 0.1 s. The results of these experiments showing the data averaged over time periods of one minute for various locations in the Lincoln Tunnel were reported by us previously[8] and are shown in Figs. 7 through 9. While these do not reflect a behavior exactly like that suggested by fluid dynamics there are noteworthy similarities that proved of assistance in estimating the maximum flows obtainable at non-bottleneck locations. This kind of knowledge is of obvious importance to the New York Port Authority in determining how much expense to incur in improving bottleneck areas of its tunnels.

References p. 192

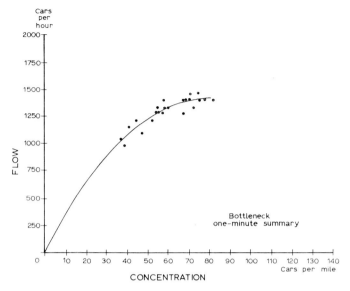

Fig. 7. Flow–concentration behavior of bottleneck. One-minute summary.

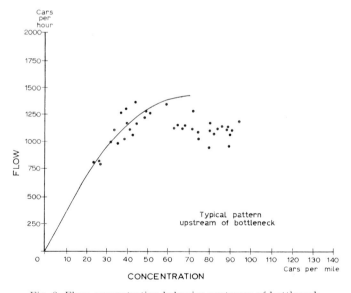

Fig. 8. Flow–concentration behavior upstream of bottleneck.

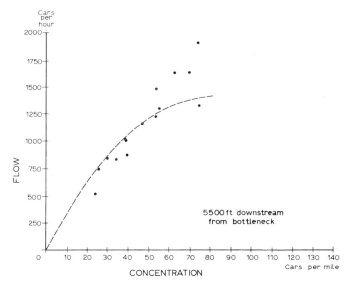

Fig. 9. Flow–concentration behavior 5500 ft downstream from bottleneck.

GREENBERG'S MODEL

The usefulness of the fluid flow theory was increased by the work of New York Port Authority staff member GREENBERG[9] in giving the flow–density relationship a precise mathematical description which agreed well with observations. Greenberg postulated the classical equation of motion for a one-dimensional fluid in addition to the continuity equation and deduced therefrom a steady state relationship of $q = ck \ln k_j/k$ as a steady state condition, where q = flow in vehicles per unit time, c = the stream velocity at maximum flow, k = density or concentration in vehicles per unit length and k_j = maximum density. How well this model fits certain data is shown in Fig. 10.

When GREENBERG's model is normalized and plotted, it has the form shown in Fig. 11. Maximum flow occurs at 37% of jam density and data taken under noncongested flow could be used to estimate the maximum flow value.

Following investigation of the fluid flow model for measuring capacity at a point, experiments using this method were subsequently concentrated on the Holland Tunnel South Tube. This tube has shown rather poor flow performance compared to other tubes and is the most frequently congested of all New York Port Authority tubes. In the face of daily excesses of traffic demand over tube capacity, the peak-hour flows in both lanes achieved values around 1700 vehicles per hour in the morning as compared with 1800 for the North Tube of this tunnel with similar traffic composition in the afternoon; and values of around 2000 vehicles per hour in evening peaks as compared with 2150 for the Lincoln Tunnel with similar traffic composition.

Measurements were made with the Simplex Productograph at a number of points in both lanes of this tube and also by means of stop-watch counts of maximum one-minute flows. The resulting profiles of maximum flows, given in Fig. 12, show the bottleneck to be located at the bottom of the tunnel upgrade and to

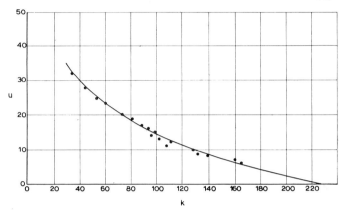

Fig. 10. GREENBERG's model of speed–density behavior compared with observations in the Lincoln tunnel.

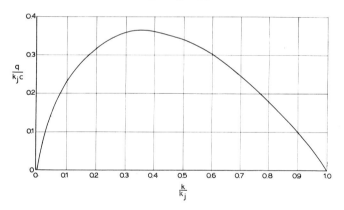

Fig. 11. GREENBERG's model normalized for flow and density.

average about 1250 vehicles per hour in the fast or near lane and 675 vehicles per hour in the far or slow lane. The top curves should be noted in particular.

CONSTANT SPEED PLATOON EXPERIMENTS

The Productograph data, analyzed on flow–density coordinates, were quite effective in describing behavior of traffic at a point but proved to be a very time-

References p. 192

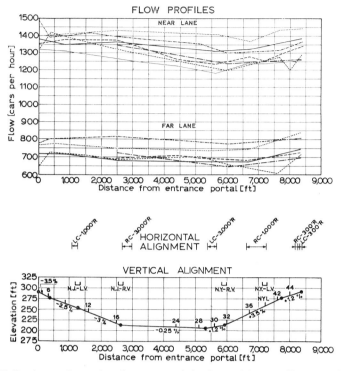

Fig. 12. Holland tunnel, south tube, near and far lanes. Flow profiles and alignment.

consuming method for obtaining flow profiles portal to portal. In working with them, one observed that capacity flow occurred at speeds of about 20 mph or somewhat less. If traffic could be regulated to flow at this speed, one could surmise that capacity flow would be observed directly and quickly at any number of locations. This possibility led to a series of experiments using speed controlled platoons having a New York Port Authority vehicle at the head and another at the tail of an 11-car platoon which included 9 patron vehicles selected at random. The driver of the lead vehicle was instructed to maintain a speed of 20–25 mph. Each New York Port Authority vehicle carried an observer who recorded the exact time his vehicle passed traffic signal control stations. The time difference between the two New York Port Authority vehicles in passing any point gave the sum of 10 time headways, from which could be deduced the flow rate within the platoon.

When the flow rates within the platoons were plotted against tunnel locations for 4 typical runs the flow profiles shown in Fig. 13 resulted. The numbers along the curves, give the computed speed of the lead car above and that of the tail car below. An appreciable range of flow values were obtained with a variety of patterns. Two runs, cycles 5 and 6, started at low values of 750 vehicles per hour and

References p. 192

rose slowly along a monotonic curve to values around 1500 and 1700 vehicles per hour. Both then showed dips followed by rises, cycle 6 showing the severest effect. The slow initial rise in flow resulted from the condition that the platoon was not fully formed when the run started. Cycle 7 starts between 1500 and 1750, then declines down to about 1100 at the exit. Cycle 8 starts around 1800 and stays at about that level.

All the values of these runs which met certain arbitrary criteria of optimum or super-optimum speeds were averaged together to yield the average flow curve

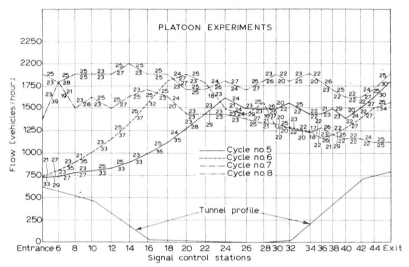

Fig. 13. Holland tunnel, south tube, fast lane. Flow profiles developed by 11-car platoons

shown in Fig. 14. Also shown is the average of the productograph and other experiments with continuous flow. The criteria used in selecting platoon values to average were: speeds of the lead vehicle between 20–30 mph; a speed change of no more than 5 mph between observation points; and complete platoon formation plus two stations. The averages of the platoon experiments appear to reflect the best flow values obtainable at steady speeds between 20–30 mph. These are much higher than the peak flows regularly produced in this lane without changing driver or tunnel characteristics. Also of interest are the two minimal values and their locations which are displaced some distance downstream from grade-change points.

FLOW WAVE EXPERIMENTS

To investigate the interacting conditions which prevented in continuous streams the flow rates found in the platoons at speeds of 20–30 mph, experiments involving simultaneous measurements at a number of locations were conducted. This method

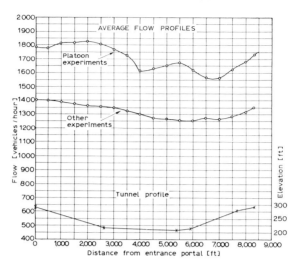

Fig. 14. Holland tunnel, south tube, fast lane. Average flow profiles by platoons and other experiments.

uses observers with synchronized stop watches who record in 15-second time boxes the exact sequence of vehicles passing them within the time period. A sample of such data would appear as shown in Table III.

TABLE III

SAMPLE DATA SHEET FOR FLOW WAVE EXPERIMENTS

Minute	0–15″	15–30″	Number of last vehicle	30–45″	45–60″	Number of last vehicle
60	CCTC	CCBCC	9	CCCCT	CCBCCC	20
01	CCCCT	CCCBC	30	CCCCC	CBCCCT	41
02	BCCCT	CCCCTC	52	CTCCC	BCCT	61
03	CCCCCC	CCCCBC	73	CCCB	CCC	80

Since lane changing is prohibited in the tunnel, the same sequence of vehicles should occur at all points but at different times. This permits vehicles to be numbered in sequence and to be identified when they pass each observer. These data could then be plotted to indicate the general behavior of all vehicles passing through the tunnel over a period of time in the manner shown in Fig. 15. Here the successive vehicles are identified by ordinates and their times of passage by abscissae. The locus of such points form a curve whose slope represents flow in vehicles per minute passing the observer.

In effect, this summary permits the analyst to take the roof off the tunnel and observe flow behavior throughout the tunnel. When one studies the resultant curves for several observers the propagation of certain flow waves becomes evident.

References p. 192

These are shown by dashed lines. The space and time origins of these waves are apparent and the propagation velocities forward and backward in space can be measured at the values shown. For comparison with these values, kinematic wave theory would estimate the backward waves at about 10 mph for a change in flow from 1400 to zero and a change in concentration from about 70 to 210 or for any linear interpolation of values in between. Values of such magnitude are found in

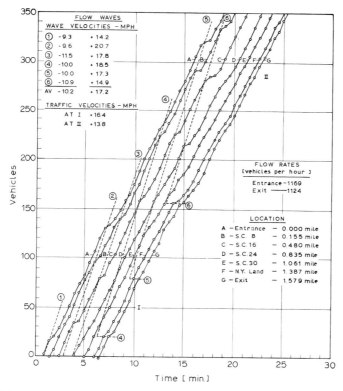

Fig. 15. Holland tunnel, south tube, fast lane. Flow wave experiments.

the data. Also for comparison one could estimate the forward wave velocities using a Δq of 1200 and a Δk of 70 to be about 17 mph. Again the results are reasonably consistent with the requirement of continuity and observation. Fig. 15 reflects a period of degeneration when congestion was increasing and flow decreasing, as indicated by a drop in speed from 16.4 down to 13.8 mph.

GAP EXPERIMENTS

A qualitative analysis of the productograph, platoon and wave experiments revealed that high traffic flows for time periods of an hour or less were associated

References p. 192

with speeds of around 20 mph, concentrations of about 70 vehicles per mile, and rates of flow of 1400 vehicles per hour or better in the Holland Tunnel South Tube fast lane. Further qualitative and quantitative analysis revealed the existence of gaps in the traffic stream of 4 seconds and more when these maximum flows were observed. This was the case for the high profiles in Fig. 12. Also there were no shock waves present. From these facts, GREENBERG and other staff members concluded that congested flow could be improved by the deliberate creation of gaps in a traffic stream. Further, they estimated that flows of 1320 could possibly be obtained by introducing these gaps at the entrance to the Holland Tunnel South Tube near lane. This conclusion was confirmed by HELLY[6] with his computer simulation studies.

Experiments have been conducted to test these predictions and the results have been encouraging. Out of twenty-five experimental days during June and July of this year, traffic was controlled on twelve days and uncontrolled on thirteen days. The type of control imposed was one of introducing gaps into the traffic stream whenever the flow became excessive. In shooting for a flow of 1320, the criterion of control used was to count and time the flow in two-minute intervals. Whenever the inflow at the entrance reached 44 vehicles before the two minutes had expired, a police officer under instructions from the field experimenters would stop traffic for a few seconds until the two minutes had elapsed. Then the count would be restarted for the next two minutes. With perfect control the hourly flow would be $30 \times 44 = 1320$ vehicles per hour.

As with many experiments, perfect control was not possible. Various difficulties arose including some lane changing by patrons, the filling of gaps by police officers inside the tunnel, and insufficient temporary output from the toll lanes to provide 44 vehicles every two minutes. Despite these difficulties, it was possible in some cases to achieve flows of around 1300 vehicles per hour for periods of time of approximately one hour. The average for twelve test days was 1248, as compared with an average of 1176 for non-test days when traffic was permitted to flow in the usual way. Thus the days the gaps were introduced averaged 72 vehicles per hour more than the other days. This increase of 6% is considered quite significant.

A better illustration of the effect on the traffic stream of introducing gaps is given in Fig. 16. When this time period began, traffic was in transition from a state of congested flow with associated shock waves to a more fluid flow without them. To establish complete transition and a condition of more fluid flow very large gaps of 30 seconds or more were introduced after each appearance of a shock wave at the entrance portal. After two more such efforts, the improved flow condition was established and it continued off-scale to be maintained by the introduction of the small gaps as previously described at two-minute intervals.

In this illustration, the results were even better than predicted. The flow rate at the beginning of the period was about 1200 vehicles per hour, with a stream velocity of 15.1 mph and a density of 80 vehicles per mile. At the end of the period the flow rate had been increased to 1360 vehicles per hour at a velocity of 19.4 mph and a density of 70 vehicles per mile.

While the direct increase in the outflow rate was the most important result of the gap experiments, there were a number of ancillary benefits as well. Because of the increase in stream velocity of nearly 30%, the transit time for each patron through the tunnel (length 1.58 miles) was reduced by one minute or more. When one considers the number of patron minutes thus saved in say a thousand peak

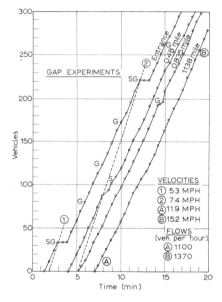

Fig. 16. Holland tunnel, south tube, fast lane. Gap experiments.

hours per year by over a thousand patrons per hour, it would add up to more than ten man years. Possibly an even more important benefit is the reduced exposure of vehicles to breakdowns. These breakdowns are responsible for an average lost capacity of about seventy-five vehicles per peak hour. A reduction in this loss is anticipated for three reasons: (1) there are fewer vehicles in the tunnel at any one time, (2) each vehicle is in the tunnel for a shorter period of time, (3) and probably of greatest importance, the stop-and-start driving caused by heavy congestion is eliminated. Another benefit to the motorist is the improved quality of the flow resulting from fewer speed changes. Most suggestions for quality indexes for traffic flow integrate speed changes or use average acceleration in the numerator, but independently of any arbitrary quality index the reduction of speed changes quite obviously improves quality.

References p. 192

This method of improving flow will be the subject of further experimentation at the New York Port Authority tunnels, by providing for the automatic introduction of slight gaps in the stream when inflow becomes excessive. In addition, research is being directed at the structure of platoons of vehicles between gaps in order to determine optimum structures, and changes in structure as they pass through the tunnel. Also research is contemplated on the saturation process which occurs at bottleneck locations, in order to find a more explicit model of traffic behavior which includes roadway, vehicle and driver parameters.

In summary, we believe that significant knowledge and understanding of traffic flow has been derived from our experiments; that we now have enough insight into the processes taking place to bring about small but important improvements in tunnel flow like that described herein. Before closing we should perhaps comment on a seeming anomaly or contradiction between the two experiments cited whereby increased peak flows were achieved. At the Lincoln Tunnel the increase was obtained by eliminating gaps at the entrance; at the Holland Tunnel the increase was obtained by creating gaps at the entrance. How does this happen?

Actually, the real cause of better flow was probably the same in both cases, namely that of preventing short period flows greater than the tunnel lane could carry. It appears that the original conclusion that gaps were stealing capacity could be in error; the higher flows between gaps achieved with officer control might have been creating jam-ups at the bottom of the downgrade.

REFERENCES

1 R. I. STRICKLAND, Traffic Operation at Vehicular Tunnels, *Proc. Highway Research Board*, *33* (1954) 395–404.

2 E. S. OLCOTT, The Influence of Vehicular Speed and Spacing on Tunnel Capacity, *Operat. Research*, *3* (1955) 147–167.

3 *Tunnel Traffic Capacity Reports II and IV*, Port of New York Authority, October 1956, January 1957.

4 T. W. FORBES, M. J. ZAGORSKI, E. L. HOLSHOUSER AND W. A. DETERLINE, Measurement of Driver Reactions to Tunnel Conditions, *Proc. Highway Research Board*, *37* (1958) 345–357.

5 D. C. GAZIS, R. HERMAN AND R. B. POTTS, Car-Following Theory of Steady-State Traffic Flow, *Operat. Research*, *7* (1959) 499–505.

6 W. HELLY, Dynamics of Single-Lane Vehicular Traffic Flow, *Research Rept no. 2, Center Operat. Research*, Mass. Inst. Technol., 1959, and in R. HERMAN (ed.), *Proc. Symposium on Traffic Flow, Detroit 1959*, Elsevier, Amsterdam, 1960, p. 207.

7 M. J. LIGHTHILL AND G. B. WHITHAM, On Kinematic Waves II, A Theory of Traffic Flow on Long Crowded Roads, *Proc. Roy. Soc. London, A 229* (1955), 317–345.

8 L. C. EDIE AND R. S. FOOTE, Traffic Flow in Tunnels, *Proc. Highway Research Board*, *37* (1958) 334–344.

9 H. GREENBERG, An Analysis of Traffic Flow, *Operat. Research*, *7* (1959) 79–85.

A Theory of Traffic Flow in Tunnels

G. F. NEWELL

Brown University, Providence, Rhode Island

ABSTRACT

If cars are constrained to follow one another with no passing, we postulate that each car has its own velocity–headway relation and consequently the motion of a series of cars is described by a set of nonlinear differential equations. Simple approximate solutions of these equations can be constructed provided the velocities vary slowly enough, even though different cars have different velocity–headway relations. These approximations are used to describe the flow in tunnels with particular emphasis on the consequences of variations in the maximum desired speeds of the cars.

INTRODUCTION

We consider here a model for traffic flow with no passing of cars which is an amalgamation of the car following theories studied previously by REUSHEL[1], PIPES[2], CHANDLER, HERMAN AND MONTROLL[3], CHOW[4], KOMETANI AND SASAKI[5], HERMAN *et al.*[6], GAZIS, HERMAN AND POTTS[7], NEWELL[8] and HELLY[9]. Some approximate solutions for this model will then be used to examine certain features of the traffic flow in tunnels.

REUSHEL and PIPES proposed a theory in which the velocity $U_k(t)$ of any kth car at time t is some linear function of the instantaneous distance $D_k(t)$ from the car ahead, *i.e.*

$$U_k(t) = \lambda_k D_k(t) - \alpha_k \tag{1}$$

where α_k and λ_k constants for any kth car. CHANDLER *et al.* assumed that the acceleration of any kth car at time t is a linear function of the spacing D_k and the relative velocity $\mathrm{d}D_k/\mathrm{d}t$ evaluated at some earlier time, $t - \Delta_k$. Their experiments showed, however, that there was little correlation between the acceleration and spacing. If one neglects this, their equations become perfect differentials with respect to the time which can be integrated to give a linear relation between $U_k(t)$ and $D_k(t - \Delta_k)$, *i.e.*

$$U_k(t) = \lambda_k D_k (t - \Delta_k) - \alpha_k \tag{2}$$

References p. 205/206

in which λ_k, α_k and Δ_k are positive constants. Eq. (1) is, therefore, a special case of Eq. (2) with $\Delta_k = 0$. KOMETANI AND SASAKI also independently proposed an equation of the form of Eq. (2). This relation is shown as curve 1 of Fig. 1a.

These equations should give a reasonable description of the motion for a sequence of cars when fluctuations in the spacings D_k are small. They are ideally suited to the investigation of stability with respect to small perturbations but are not suited to the study of phenomena in which wide variations in D_k occur or situations in which D_k is so large that the cars are nearly free. For a sequence of equally spaced, identical cars, $D_k = D$, $U_k = U$, $\lambda_k = \lambda$ etc. for all k, one finds

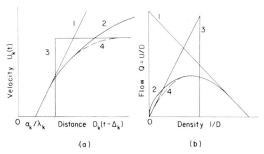

Fig. 1. Several types of approximate theoretical relations between velocity and spacing are shown in (a) along with the resulting flow–density relations (b).

that Eq. (1) or (2) gives a relation between the steady flow $Q = U/D$ and the density $1/D$

$$Q = \lambda - \alpha(1/D) \tag{3}$$

which fails to show a maximum Q for $1/D > 0$ as is observed experimentally. Eq. (3) is shown by curve 1 in Fig. 1b.

To correct this limitation on the theory, GAZIS, HERMAN AND POTTS proposed a theory in which the velocity and spacing were related through an equation of the form

$$U_k(t) = \lambda'_k \log [D_k(t - \Delta_k)] - \alpha'_k \tag{4}$$

with λ'_k, α'_k and Δ_k constants. For small changes in D_k, one can approximate this by a linear relation such as Eq. (2) but for large variations in the D_k the equation leads to a steady state flow–density relation that agrees with the one proposed by GREENBERG[10] and which has been found to give remarkably good agreement with experimental data over a wide range of densities. Eq. (4) is shown by curve 2 in Fig. 1a and the corresponding Q vs $1/D$ relation is shown by curve 2 of Fig. 1b.

In the model proposed by NEWELL, the velocity vs spacing relation was assumed to be rectangular such as shown by curve 3 of Fig. 1a and leads to the Q vs $1/D$ relation shown by curve 3 of Fig. 1b. This is a very crude model but was used primarily to illustrate the consequences of imposing an upper bound on the velocities of cars, a necessary feature which is still lacking in Eq. (4).

All of these theories have the common feature that the velocity is considered to be some function of the spacing, observed either at the same time or at some earlier time *i.e.*

$$U_k(t) = G_k \{D_k (t - \Delta_k)\} \tag{5}$$

for some particular function G_k. To amalgamate all of the desirable features of the above models into one, we should take for G_k a function similar to that shown by curve 4 of Fig. 1a, which differs from curve 2 mainly in that U_k is bounded by the free speed V_k of the kth car and gives a flow–density relation with a finite slope as $1/D \to 0$. If we wish to consider the formation of queues created by slow cars, then we must also allow G_k to vary from car to car, so as to give different values of V_k.

Another special form for G_k which was proposed by NEWELL and HELLY but which has not been studied in any detail is obtained by combining curves 1 and 3 of Fig. 1a and b. We let

$$U_k(t) = \begin{cases} 0 & \text{for } 0 \le D_k (t - \Delta_k) \le \alpha_k/\lambda_k \\ \lambda_k D_k (t - \Delta_k) - \alpha_k & \text{for } \alpha_k/\lambda_k \le D_k (t - \Delta_k) \le (V_k + \alpha_k)/\lambda_k \\ V_k & \text{for } (V_k + \alpha_k)/\lambda_k \le D_k (t - \Delta_k) \end{cases} \tag{6}$$

This gives the triangular Q *vs* $1/D$ relation obtained by using curve 3 of Fig. 1b for small densities and curve 1 for the large densities.

For any specified functions G_k, Eq. (5) represents a set of nonlinear differential-difference equations, the complete formal solution of which one could construct using some iterative schemes. Except in very special cases, however, most of which have already been studied, the exact solutions of Eq. (5) are too complicated to be of much practical value. It is therefore advantageous to consider some approximate solutions.

We shall be primarily interested here in properties of the trajectories of cars as seen on a scale of time large compared with the two basic time constants of the theory, one of which is Δ_k and the other of which is the reciprocal slope of G_k at some typical value of D_k (λ_k^{-1} in Eq. (2)). Values of these time constants obtained from various sources generally lie in the range of 1 to 3 seconds. We also assume that these constants are such as to guarantee stability in the sense discussed by CHANDLER *et al.*[3], KOMETANI AND SASAKI[5] and HERMAN *et al.*[6]

SOME EXAMPLES

Suppose as illustrated in Fig. 2, a $(k-1)$th car travels at some velocity U for $t < 0$ and at another velocity U' for $t > 0$. Suppose also that for $t < 0$, a kth car with $V_k > U$ follows the $(k-1)$th car at the same constant velocity U and at a distance $D = G_k^{-1}(U)$ consistent with Eq. (5). If $V_k > U'$, we expect that the kth car will, for $t > 0$, try to follow the leading car and will eventually achieve the constant velocity U' and a spacing $D' = G_k^{-1}(U')$. KOMETANI AND SASAKI[5], and HERMAN *et al.*[6] have shown that if the trailing car faithfully follows the motion described by Eq. (2) and if $\lambda_k \Delta_k < \pi/2$, then the velocity of the kth car can be represented by U' plus a transient that decays exponentially in time. If,

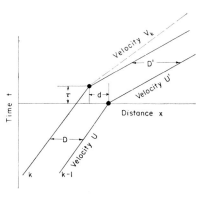

Fig. 2. Piecewise linear approximations to the trajectories of a $(k-1)$th and a kth car are shown for an example in which the $(k-1)$th car changes velocity from U to U'. The solid curve for car k represents the trajectory if U' is less than the free speed V_k of the kth car and the broken line results if $U < V_k < U'$.

however, we look at these trajectories on a coarse time scale, we can disregard the details of the transient motion which lasts only a time of order Δ_k or λ_k^{-1} and replace the true trajectory by a piecewise linear curve represented by its asymptotes, as shown in Fig. 2 by the solid line. (Note that in Fig. 2 $V_k < U'$.)

If we let τ denote the time lag between the discontinuities of the two trajectories shown in Fig. 2, d the special lag and $v = -d/\tau$ the effective velocity of propagation of the disturbance from the $(k-1)$th to the kth car, then one can easily deduce from the geometry of Fig. 2 that

$$\tau = \frac{D-D'}{U-U'} \qquad \text{and} \qquad d = D - U\frac{(D-D')}{(U-U')} \qquad (7)$$

The values of U, D and U', D' represent two points on the U_k vs D_k curve for the kth car and if we draw a straight line through these two points, τ is the slope

References p. 205/206

of the line, d the intercept with the D_k axis, and v the intercept with the U_k axis as shown in Fig. 3a. If we plot $Q_k = U_k/D_k$ vs $1/D_k$ as in Fig. 3b to represent the flow vs density relation for the kth car, then the points $Q = U/D$, $1/D$ and $Q' = U'/D'$, $1/D'$ lie on this curve. The line passing through these two points will have a slope v; the intercept on the Q_k axis is $1/\tau$; and the intercept on the $1/D_k$ axis is $1/d$. Clearly, the velocity v is the individual car equivalent of the shock velocity or, for infinitesimal velocity changes, the wave velocity in the continuum theories of LIGHTHILL AND WHITHAM[11] and RICHARD[12].

If we use Eq. (6) to represent G_k, then there are only two possible wave velocities, V_k if the car is free and $-\alpha_k$ otherwise. Also, for U and $U' < V_k$, τ and d are independent of the velocities U and U'; $\tau = \lambda_k^{-1}$ and $d = \alpha_k/\lambda_k$.

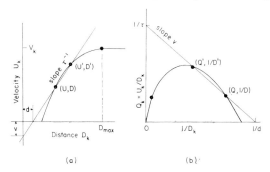

Fig. 3. The initial and final velocities and spacings (U, D) and (U', D') of Fig. 2 are represented by two points on the velocity-spacing curve (a) and the flow-density curve (b) for the kth car. A straight line through these two points determines the time lag τ and special lag d of Fig. 2 and the wave velocity $v = d/\tau$.

Suppose now that the lead car has the same motion as described above but that $U < V_k < U'$. The trailing car, after following at the prescribed distance $D = G_k^{-1}(U)$ for $t < 0$, will eventually achieve its free speed V_k and the gap between the two cars will continue to grow. In Fig. 2, the slope of the final trajectory is therefore V_k^{-1} as shown by the broken line and the two straight line asymptotes of the trajectory intersect. It is apparently not possible however to determine where they intersect without solving the differential equation, except in two extreme cases.

If we take $U' = V_k$ and assume that in the U_k vs D_k curve of Fig. 3a there is a distance D_{max} beyond which $U_k = V_k$ then by letting the point U', D' approach V_k, D_{max} we can obtain limiting values for v, d and τ. If U is also close to V_k, then the distance d will probably be negative and perhaps quite large, even infinite if at D_{max}, $G_k(D_k)$ has continuous slope with value zero. In the special case represented by Eq. (6), however, in which v, τ and d are independent of U and U' for U and $U' < V_k$, the limiting values are still $-\alpha_k$, λ_k^{-1} and α_k/λ_k, respectively.

References p. 205/206

If on the other hand we take $U' = \infty$, the trailing car maintains the velocity U until time $\varDelta_k$ but is then free to travel with the velocity V_k immediately. For any realistic value of $\varDelta_k$, this will give a positive value of d and a backward velocity for the discontinuity that is larger even than the fastest wave. Although it is not physically realistic to let U' be infinite, this might serve to represent the sudden removal of the $(k-1)$th car as an obstacle to the free motion of the kth car.

Fig. 4 illustrates an example in which a $(k-1)$th car traveling with a constant velocity U is overtaken by a kth car traveling with a velocity $V_k > U$. The trailing car starts to decelerate at a time $\varDelta_k$ after the spacing has decreased to the value D_{max} but will eventually travel at a velocity U and at a distance $D = G_k^{-1}(U)$ behind the $(k-1)$th car. If we again disregard the transients, we can approximate the trajectory by the extrapolation of its two asymptotes as shown in Fig. 4.

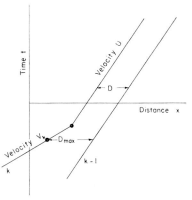

Fig. 4. A piecewise linear approximation for the trajectory of a kth car when it overtakes a $(k-1)$th car moving at a velocity $U < V_k$.

The above examples furnish a prescription for obtaining piecewise linear approximations to the trajectories of any sequence of cars. If a first car has a piecewise linear trajectory, one can obtain a piecewise linear trajectory for a second car by applying the above rules separately to each discontinuity in velocity of the first car. By iteration one obtains corresponding trajectories for subsequent cars. There are, of course, many dangers in this scheme of approximation. If the discontinuities occur too close together along a trajectory one never escapes the influence of the transients; if one tries to iterate a large number of times, the errors of approximation may accumulate to serious proportions; if two waves travel with different speeds, they may overtake one another and cause something analogous to a shock; etc. In estimating the seriousness of these things, however, one should also notice that a long sequence of cars is likely to form platoons and that a disturbance will terminate whenever it reaches a large gap in the traffic. Also, whereas the duration of a disturbance may grow as it propagates, the duration is not expected to increase as rapidly as the distance it has traveled.

A SIMPLE APPLICATION TO TUNNEL TRAFFIC

We now consider the traffic flow in a homogeneous tunnel following the same type of arguments used in the past[8] but postulate that G_k is given by Eq. (6) and the approximation scheme of the last section is applicable. For simplicity, we also

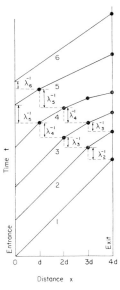

Fig. 5. A sequence of trajectories for cars passing through a hypothetical tunnel is shown. When relatively slow cars (cars 1, 2) leave the exit, an accaleration wave propagates back into the tunnel from the exit. Gaps form when velocities exceed the desired speed of some cars (cars 5, 6).

assume that $d = \alpha_k/\lambda_k$, which is also the minimum spacing between cars corresponding to zero velocity, is the same for all cars.

We assume that there is a queue of cars waiting at the entrance of the tunnel and each car enters at the earliest time allowed by the theory, a time λ_k^{-1} after its predecessor has reached a distance d from the entrance. We imagine that the exit is free of congestion and that when a car leaves the exit the car behind it is permitted to travel at its free speed by the time it reaches a distance d from the exit, if it has not already done so earlier.

A typical sequence of trajectories, analogous to those shown elsewhere[8], is given in Fig. 5 for a tunnel with length Nd with N integer. A car 1 enters an empty tunnel and travels with a constant velocity V_1 throughout the tunnel. A second car with $V_2 > V_1$ follows car I at the prescribed distance $(V_1 + \alpha_2)/\lambda_2$ until it reaches a point at a distance d from the exit where it changes velocity to its desired speed V_2. A third car with $V_3 > V_2$ changes speed from V_1 to V_2 at a distance $2d$ from the exit in response to the velocity change of car 2 and again from V_2 to V_3 at a distance d from the exit. A fourth car with $V_4 > V_3$ changes speed at each multiple of the distance d. The fifth car with $V_2 < V_5 < V_3$ starts with velocity V_2 but at a time λ_5^{-1} after the fourth car changes velocity from V_2 to V_3, car 5 changes velocity from V_2 to V_5 and keeps the velocity V_5 throughout the remaining length of the tunnel allowing a gap to grow between the fourth and fifth cars. The sixth car is another slow car with $V_6 < V_5$ and it travels with its free speed throughout the tunnel.

If we let U_{kj} be the velocity of the kth car in the jth interval $(j-1)d < x < jd$, then

$$U_{kj} = \min (V_k, V_{k-1}, \ldots, V_{k+j-N}) \tag{8}$$

The time difference between the arrivals of the kth and $(k-1)$th cars at the point $x = jd$ is

$$T_{kj} = \frac{1}{\lambda_k} + \frac{d}{U_{k-1,1}} + d \sum_{l=1}^{j} \left(\frac{1}{U_{k,l}} - \frac{1}{U_{k-1,l}} \right) \tag{9}$$

and the spacial gap D_{kj} between the kth and $(k+1)$th cars when the kth car reaches $x = jd$ can be obtained from T_{kj} through the relation

$$D_{k-1,j} = T_{kj} U_{kj} \tag{10}$$

If we consider the $\{V_k\}$ to be independent and identically distributed random variables with a distribution function

$$F(v) = \Pr \{V_k \leq v\}$$

then the U_{kj} have distribution functions

$$F_j(u) = \Pr \{U_{kj} \leq u\} = 1 - [1 - F(u)]^{N-j+1} \tag{11}$$

and the average velocity at $x = jd$, for $N - j + 1 \gg 1$, is approximately

$$E(U_{kj}) \sim F^{-1}[1/(N - j + 1)] \tag{12}$$

where $E(U_{kj})$ is the expectation value.

This is exactly the same result as derived by us before[8] except for some differences in the interpretation of the parameters, and leads to the conclusion that the velocities of cars increase as they travel from the entrance to the exit.

References p. 205/206

If we consider the $\{\lambda_k\}$ to be identically distributed random variables and let $\lambda^{-1} = E(\lambda_k^{-1})$, the capacity of the tunnel is given by

$$q^{-1} = E\,(T_{kj}) = E\,(\lambda_k^{-1}) + dE\,(U_{k-1,\,1}^{-1}) \sim \frac{\text{I}}{\lambda} + \frac{d}{F^{-1}\,(\text{I}/N)} \tag{13}$$

which is a monotonically decreasing function of N, the length of the tunnel.

The distribution for the lengths of the gaps between cars is more complicated here than in our former study[8] because we must now take notice of the fact that even for two consecutive cars having exactly the same velocity, the gap between them is a random variable depending upon the values of λ_k and U_k. We can, however, still give a crude description of the average size of the gaps in front of freely moving cars using arguments similar to those formerly used.[8]

A $(k + \text{I})$th car will travel at its desired speed V_{k+1} before it reaches the point $x = jd$ if and only if $V_{k+1} < V_k, \ldots,$ and V_{k+j-N}. The probability for this is $\text{I}/(N - j + 2)$. Thus, for any random variable Y

$$E\,(Y) = \left[\text{I} - \frac{\text{I}}{(N - j + 2)}\right] E_1\,(Y) + \frac{\text{I}}{(N - j + 2)} E_2\,(Y)$$

in which E_1 and E_2 denote conditional expectations given that $U_{k+1,j} < V_{k+1}$ [the $(k + \text{I})$th car is constrained by the kth car prior to the time the $(k + \text{I})$th car reaches $x = jd$] and $U_{k+1,j} = V_{k+1}$ [the $(k + \text{I})$th car is free by the time it reaches $x = jd$], respectively.

The random variable $Y = T_{k+1,j} - \lambda_{k+1}^{-1} - d/U_{k+1,j}$ represents the actual time gap between cars at $x = jd$ relative to what it would be if the $(k + \text{I})$th car were traveling with the velocity $U_{k+1,j}$ and had a gap consistent with Eq. (2). If the $(k + \text{I})$th car is constrained by the kth car then $Y = 0$, so $E_1(Y) = 0$ and

$$E_2\,(T_{k+1,j} - \lambda_{k+1}^{-1} - d/U_{k+1,j}) = (N - j + 2)E\,(T_{k+1,j} - \lambda_{k+1}^{-1} - d/U_{k+1,j})$$

The right hand side can be evaluated using Eq. (13) and $T_{k+1,j}$ can be expressed in terms of D_{kj} using Eq. (10). We thus obtain

$$E_2\left(\frac{D_{kj} - U_{k+1,j}\,\lambda_{k+1}^{-1} - d}{U_{k+1,j}\,E\,(\text{I}/U_{k+1,j})}\right) = (N - j + 2)\,d\left\{\frac{E\,(\text{I}/U_{k+1,1})}{E\,(\text{I}/U_{k+1,j})} - \text{I}\right\} \tag{14}$$

which is the analogue of Eq. (13) in our former paper[8].

The left side of Eq. (14) is a weighted average of the length of the gap in excess of what Eq. (2) would predict, conditional that there be an excess gap. From this we estimate, however,

$$E_2[D_{kj} - U_{k+1,\,j}\,\lambda_{k+1}^{-1} - d] = d\,(N - j + 2)\left\{\frac{F^{-1}\,[\text{I}/(N - j + \text{I})]}{F^{-1}\,[\text{I}/N]} - \text{I}\right\} \tag{15}$$

which is the generalization of Eq. (14) of our former paper. This new formula, however, leads to essentially the same conclusions as described before[8].

Generally we find here that all the qualitative conclusions discussed] on the basis of a very crude model of car following are still valid when we consider a more refined model.

SUBCAPACITY FLOW

In the last section we investigated only what happens if cars enter a tunnel as quickly as the theory will allow, *i.e.* the kth car enters the tunnel at a time λ_k^{-1} after the $(k-1)$th car reaches the point $x = d$. Here we will give a very crude description of what should happen if cars enter later than the minimum time, *i.e.* at a time $t_k + \lambda_k^{-1}$ after the $(k-1)$th car reaches d with $t_k > 0$.

Whereas at one extreme, $t_k = 0$ for all k, we have the case treated above, at the other extreme, t_k arbitrarily large, each car enters an empty tunnel. If, in the latter case, the tunnel is uniform then each car travels at its free speed throughout and the average velocity is $E(V_k)$ everywhere. Between these two extremes we find that a typical car travels at its free speed only until it overtakes another car. When this happens the car is forced to reduce its speed to that of its predecessor which in turn may later be forced to decrease its speed as it overtakes still other cars. The velocity of any car is therefore monotonically nonincreasing until the car for the first time becomes part of a queue the front end of which has reached the exit. Hereafter there is no longer any possibility for this queue to overtake other slower queues but instead the velocities of cars in the queue may now increase whenever a slow car leaves the exit. We therefore, expect that if $t_k > 0$, the average velocity of cars will be approximately $E(V_k)$ at both the entrance and the exit and have a single minimum at some interior point.

If cars enter a tunnel at almost the minimum time, a typical kth car is likely to overtake its predecessor at a point very close to the entrance. If in fact this should happen before the kth car has reached a distance d, the $(k+1)$th car will not yet have entered the tunnel and, therefore, will experience no reaction to the late entrance of its predecessor. If this should be true for most cars, the only effect of the late arrivals would be that all entrance times are displaced somewhat but the time difference between entrances remains nearly unchanged and consequently also the flow. The average velocity would then be relatively high at the entrance but decrease very quickly to a value comparable with the entrance speed deduced in the preceding section and, thereafter, show nearly the same increase described in this section. The trajectories would also look very much like those in Fig. 5 except that as one approaches the entrance from the interior, most trajectories are cut off by lines of smaller slope which then intercept the line $x = 0$ at a point somewhat later in time.

References p. 205/206

Actually there will be a slight decrease in flow due to these late arrivals. Even for the case $t_k = 0$, one occasionally encounters a car with a desired speed less than any of its N predecessors. If such a car enters late by an amount t_k, it will never overtake any cars and furthermore will cause all succeeding cars to enter at a time t_k later even if $t_k = 0$ for all succeeding cars. If $t_k = t_1$ for all cars, the above effect alone will cause a decrease in flow from q to approximately $q - q^2 t_1/N$. In addition, for any nonzero t_1, one will occassionally have other cars which gain on their predecessors but cannot catch them before they reach the exit. Still other cars may overtake the car ahead only after traveling a considerable distance. If the signal of such an overtaking cannot propagate back to the entrance before a slow car of the type described above enters, then the latter may be forced to enter late and propagate the time lag to its successors. A quantitative estimate of the decrease in flow due to all these effects is unfortunately difficult to compute analytically.

The above type of flow pattern is expected to occur when an 'average' car with desired speed $E(V_k)$ entering late by a time $E(t_k)$ can overtake a car with a velocity of approximately $F^{-1}(1/N)$, the average speed just inside the tunnel, within a distance of approximately d, i.e. if

$$E\,(t_k) < d\left[\frac{1}{F^{-1}\,(1/N)} - \frac{1}{E\,(V_k)}\right] \tag{16}$$

Some typical values here might be $E(V_k) \sim 30$ miles per hour, $F^{-1}(1/N) \sim 15$ miles per hour and $d \sim 40$ feet, in which case the upper limit on $E(t_k)$ would be of the order of 1 second (to within perhaps a factor of 2).

Whereas for near capacity flow, the minimum average velocity of cars occurs very close to the entrance of the tunnel, as the flow decreases the position of this minimum shifts toward the exit. At very low flow rates there is a high probability that a car will travel throughout the tunnel with no change in velocity and a small probability that it will overtake only one car somewhere in the tunnel, be slowed down and then increase speed after it has reached a point d from the exit. There is still a smaller probability that the car will join a queue of three or more cars. In this case the average velocity is nearly constant but has a shallow minimum at approximately d from the exit.

For somewhat higher rates of flow, let us imagine that a typical fast car with velocity $E(V_k) + \sigma$ (where σ equals the dispersion of V_k) enters the tunnel m cars (where $1 < m \ll N$) after a typical slow car with velocity $E(V) - \sigma$ has entered. If each of the m cars enters at a time $E(t_k)$ later than the minimum time, the fast car can potentially overtake the slow one if it can make up a time difference $mE(t_k)$ before the cars reach the exit, i.e. if

References p. 205/206

$$mE\ (t_k) \sim Nd\left\{\frac{1}{E\ (V_k) - \sigma} - \frac{1}{E\ (V_k) + \sigma}\right\} \sim \frac{2\ Nd}{[E\ (V_k)]^2} \qquad (17)$$

The fast car, which has been decreasing its speed throughout the tunnel as it overtakes slower cars, is likely to pick up speed again after receiving a signal that the slow car has left the exit, at which time the fast car is at a point md from the exit. If we solve Eq. (17) for m, the value of md therefore represents at least the order of magnitude of the distance from the exit at which the average velocity of cars has its minimum value.

This crude estimate is probably incorrect by a significant factor but it does indicate how the minimum point depends upon various other parameters. The argument is, however, limited to the case $m \ll N$ which means that

$$E\ (t_k) \gg \frac{2d\sigma}{[E\ (V)]^2} \qquad (18)$$

For typical values of d, σ, etc., the right side of Eq. (18) is again of the order of 1 second.

Fig. 6 shows how the average velocity should vary with distance according to the present model. Curve 1 results from feeding cars into the tunnel with the minimum spacing and curves 2, 3, 4 and 5 result from successively lower rates of flow with the fifth curve corresponding to zero flow.

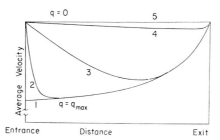

Fig. 6. The variation in average velocity between the entrance and exit of a tunnel due to a distribution of desired speeds of different cars is shown for several rates of flow starting with the maximum value (curve 1) and decreasing to zero (curve 5).

COMPARISON WITH EXPERIMENT

A considerable amount of data has been collected on traffic flow in the tunnels around New York City by the Port of New York Authority and various groups working in collaboration with them[13, 14, 15]. Although the experiments have been mainly exploratory in nature, they seem to indicate more peculiar effects than can be explained by this or any other theory proposed thus far.

The Lincoln Tunnel is the most nearly homogeneous of the tunnels in which experiments have been conducted. Here the average velocity of cars does show a fairly steady increase from the entrance to the exit during capacity operation similar to that shown by curve 1 of Fig. 6. The magnitude of this increase varies from day to day but is in the range of 50% to 150%. Some relatively crude measurements of the velocity distribution of cars made during early morning hours when the flow is so low that most cars should be traveling with nearly their free speeds indicate that about one car in 200 or so wishes to travel at about 16 miles per hour which is very close to the average entrance velocity at peak flows. The average speed at night is in excess of 30 miles per hour, a value comparable with but somewhat higher than the average exit speed at capacity flow. All of these observations are consistent with the above theory and in addition it is found that large gaps in the flow are quite noticeable near the exit.

In contrast with this, however, the flow in the Holland Tunnel, which has also been studied quite extensively, shows a pattern almost completely at variance with the present theory. The variation in the velocity from entrance to exit is similar to that shown by curve 3 of Fig. 6 but there is no evidence to indicate that this is due to any failure to feed the tunnel fast enough. The dominant feature of the flow seems to be some sort of instability that causes the tunnel to jam periodically at intervals of several minutes.

ACKNOWLEDGEMENTS

Most of the work described here was done while the author was on sabbatical leave from Brown University visiting the Stockholms Högskola as an Alfred P. Sloan Research Fellow. He would like to express his appreciation to Professor MALMQUIST, Docent DALENIUS and the staff of the Institute for Statistics for their hospitality. The work was completed under a research grant from General Motors Corporation. Mr. L. EDIE and Dr. R. HERMAN generously supplied data and reports of their work on related problems.

REFERENCES

1 A. REUSHEL, Fahrzeugbewegungen in der Kolonne, *Österr. Ingr.-Arch.*, 4 (1950) 193–215; Fahrzeugbewegungen in der Kolonne bei gleichförmig beschleunigtem oder verzögertem Leitfahrzeug, *Z. Österr. Ingr. Architekt. Vereines*, 95 (1950) 59–62, 73–77.
2 L. A. PIPES, An Operational Analysis of Traffic Dynamics, *J. Appl. Phys.*, 24 (1953) 274–281.
3 R. E. CHANDLER, R. HERMAN AND E. W. MONTROLL, Traffic Dynamics: Studies in Car Following, *Operat. Research*, 6 (1958) 165–184.

4 TSE-SUN CHOW, Operational Analysis of a Traffic-Dynamics Problem, *Operat. Research*, 6 (1958) 827–834.

5 E. KOMETANI AND T. SASAKI, On the Stability of Traffic Flow, *J. Operat. Research Japan*, 2 (1958) 11–26.

6 R. HERMAN, E. W. MONTROLL, R. B. POTTS AND R. W. ROTHERY, Traffic Dynamics: Analysis of Stability in Car Following, *Operat. Research*, 7 (1959) 86–106.

7 D. C. GAZIS, R. HERMAN AND R. B. POTTS, Car-Following Theory of Steady-State Traffic Flow, *Operat. Research*, 7 (1959) 499–505.

8 G. F. NEWELL, A Theory of Platoon Formation in Tunnel Traffic, *Operat. Research*, 7 (1959) 589–598.

9 W. HELLY, Dynamics of Single-Lane Vehicular Traffic Flow, *Research Rept. No. 2, Center Operat. Research*, Mass. Inst. of Technol., 1959, and in R. HERMAN (ed.), *Proc. Symposium on Traffic Flow, Detroit 1959*, Elsevier, Amsterdam, 1961, p. 207.

10 H. GREENBERG, An Analysis of Traffic Flow, *Operat. Research*, 7 (1959) 79–85.

11 M. J. LIGHTHILL AND G. B. WHITHAM, On Kinematic Waves II, A Theory of Traffic Flow on Long Crowded Roads, *Proc. Roy. Soc. London, A 229* (1955) 317–345.

12 P. I. RICHARDS, Shock Waves on the Highway, *Operat. Research*, 4 (1956) 42–51.

13 E. S. OLCOTT, The Influence of Vehicular Speed and Spacing on Tunnel Capacity, *Operat. Research*, 3 (1955) 147–167.

14 R. I. STRICKLAND, Traffic Operation at Vehicular Tunnels, *Proc. Highway Research Board*, 33 (1954) 395–404.

15 L. C. EDIE AND R. S. FOOTE, Traffic Flow in Tunnels, *Proc. Highway Research Board*, 37 (1958) 334–344.

Simulation of Bottlenecks in Single-Lane Traffic Flow*

W. HELLY

Bell Telephone Laboratories, New York, N. Y.

ABSTRACT

The bottleneck behavior of automobile traffic is simulated for environments where passing cannot occur and where vehicles neither enter nor leave the traffic stream. Using a 'follow-the-leader' philosophy for the individual driver, a computer model is constructed and its parameters are fitted to observational data. The model consists of a reaction time-lagged system of differential equations in which the acceleration of each driver is a function of his headway, his velocity difference from the car ahead of him, and other pertinent factors.

The model is applied to the study of bottlenecks in vehicular traffic tunnels. It is found that there are two basic types of bottlenecks, differentiated by the behavior in each of an isolated compact platoon of vehicles. In a bottleneck of the first kind, the average bottleneck time headway of the nth vehicle in such a platoon is independent of n. (n is counted backwards from the platoon leader.) In a bottleneck of the second kind, the average time headway increases with n. Under certain conditions, a bottleneck of the second kind can handle an increased traffic flow, as compared with the steady state value, if the vehicles are deliberately platooned by the introduction of substantial car-free gaps. A theoretical model of a bottleneck of the second kind is developed in the form of a Markov process.

INTRODUCTION

This paper develops a single-lane traffic flow model sufficiently detailed to be useful in studying the causes and alleviation of vehicular tunnel bottlenecks that limit flow to less than the ideal maximum which might be expected on the basis of minimal safe headways. The model is programmed for an IBM type 704 computer. The program written may be likened to the apparatus used in more orthodox experimentation. We also present experimental support for the parameters used to describe the drivers in the computer program.

We study some comparatively subtle bottlenecks caused by localized small variations in driver intent or vehicle capability. These have a substantial effect

---

* Work performed at the Massachusetts Institute of Technology, Cambridge, Mass.

References p. 237/238

on flow. While our simulation unfortunately is not firmly based on experiment, it does lead to plausible results which suggest that certain types of bottlenecks might be greatly alleviated by deliberately introducing gaps into the traffic stream. We outline experiments to determine whether this is possible. A theoretical model is advanced to describe these bottlenecks.

The Simulation Model

PRESENTATION

Let us consider the requirements for a single-lane congested traffic model. Passing will not be permitted. There are several obvious observable driver–vehicle characteristics which must be instrumented:

1. Vehicles tend to follow each other at a velocity-dependent distance.

2. Drivers have a finite non-zero reaction time.

3. Brake lights are observable and cause reactions in follower vehicles.

4. The aggregate of vehicles performs a relatively stable motion. This we deduce from the scarcity of observed accidents.

5. Drivers are limited by the accuracy of their senses and instruments in estimating headway distances and velocities. The computer model naturally has these available to any desired degree of precision. To correspond to the real driver, this precision should be marred by adding pseudo-random numbers from appropriate distributions, to the precisely determined headways and velocities.

6. A driver does not operate in a frenzy of continuous revision on his acceleration program. Rather, he makes decisions at variable time intervals, as dictated by unforeseen changes in his environment.

7. Vehicles have finite acceleration limits.

8. Each driver has some maximum velocity, perhaps dependent on location or other circumstances, which he will not exceed.

In constructing a model, we have included some features which could not be made part of the simulation because data was lacking. Nevertheless these are described because they should eventually be included in a reasonably complete formulation.

Let the successive cars be labelled 1, 2, 3, ..., $n - 1$, n, ... We assume that the nth car attemps to maintain its course along the positive x-axis by accelerating according to a rule of the form

$$\ddot{x}_n\,(t + \tau) \equiv (\ddot{x}_n)_{t+\tau} = \text{a function of some or all of the following: } x_n, x_{n-1}, \dot{x}_n, \dot{x}_{n-1},$$
$$\text{braking by cars } (n - 1), (n - 2), \text{ and similar quantities, all}$$
$$\text{evaluated at time } t.$$

Here, τ is a reaction delay time. One such rule, embodying only a velocity control, is

$$(\ddot{x}_n)_{t+\tau} = C_1 (\dot{x}_{n-1} - \dot{x}_n)_t$$

used by CHANDLER, HERMAN AND MONTROLL[1]. While this system of equations is stable if $C_1 < (2\tau)^{-1}$, collisions may occur unless the vehicles are initially spaced far apart because there is no provision for driver reactions to variations in headway. Stability here means that oscillatory behavior of the first car, $n = 1$, will induce only oscillations of decreasing magnitude in the following cars as $n \to \infty$, and not that these oscillations are small enough to prevent collisions. Thus a headway control appears necessary also.

Another rule can be derived from the California Vehicle Code which states that the nth car should attempt to maintain a desired headway

$$D = \alpha + \beta \dot{x}_n$$

where α and β are constants. PIPES[8] and HERMAN, MONTROLL, POTTS AND RO-THERY[10] have formulated this as follows

$$(\ddot{x}_n)_{t+\tau} = C_2 (x_{n-1} - x_n - D)_t = C_2 (x_{n-1} - x_n - \alpha - \beta \dot{x}_n)_t$$

As shown by HERMAN et al.[10], this rule results in system stability if $\beta > 2/C_2$, regardless of the value of τ. However, because this control fails to include the value $\dot{x}_{n-1}$, car n will collide with $(n - 1)$ unless the initial conditions are ideal. Since it is fairly evident that a real driver does note the velocity of the car ahead of him, this second control rule is also insufficient.

In our formulation, we suppose that the driver will seek to minimize both the velocity difference, $(\dot{x}_{n-1} - \dot{x}_n)$, and the difference between his actual headway and his desired headway, $(x_{n-1} - x_n - D)$. We use a simple linear combination of the two controls and add to these two terms to provide for a response to observed brake lights. Thus, if the driver has the same speed as the car in front of him and if his headway is D, he will neither accelerate nor brake. But, if either quantity is non-zero, its effect on the control acceleration is such as to reduce this quantity. The combined control equation is

$$(\ddot{x}_n)_{t+\tau} = [C_1 (\dot{x}_{n-1} - \dot{x}_n) + C_2 (x_{n-1} - x_n - D) + C_3 B_{n-1} + C_4 B_{n-2}]_t \qquad (1)$$

where
$\quad x_n = $ location of the nth car
$\quad \dot{x}_n = $ velocity of the nth car
$\quad \ddot{x}_n = $ acceleration of the nth car
$\quad t = $ time
$\quad \tau = $ driver reaction delay time
$\quad C_1 = $ velocity control parameter, $C_1 > 0$

C_2 = headway control parameter, $C_2 \geq 0$
C_3 = brake factor relative to car $n - 1$, $C_3 \leq 0$
C_4 = brake factor relative to car $n - 2$, $C_4 \leq 0$
D = the desired headway for car n.

We make

$$D = \alpha + \beta \dot{x}_n + \gamma \ddot{x}_n \tag{2b}$$

where α, β, and γ are constants. We defer discussion of this and subsequent relations to the section on parameter fitting. Here we note only that $\beta > 0$, and so D increases with velocity as expected.

$$B_{n-1} = \begin{cases} 0 \text{ if car } n - 1 \text{ is not braking} \\ 1 \text{ if car } n - 1 \text{ is braking} \end{cases}$$

$$\tag{3}$$

$$B_{n-2} = \begin{cases} 0 \text{ if car } n - 2 \text{ is not braking} \\ 1 \text{ if car } n - 2 \text{ is braking} \end{cases}$$

Car n is assumed to be braking when

$$\ddot{x}_n < F\dot{x}_n \tag{4}$$

where F = 'friction', or braking constant. Clearly, we might add more terms to cover braking by cars $n - 3$, $n - 4$, ... However, in the absence of experimental evidence on long-range observations, we keep only two such terms.

Driver observational errors are provided by adding suitably distributed random numbers with mean zero and driver dependent variance to x_{n-1} and $\dot{x}_{n-1}$ on the right side of the acceleration equation. Discrete decisions are instrumented by having driver n estimate x_{n-1} and $\dot{x}_{n-1}$ at the beginning of his run. He then assumes that the velocity of car $n - 1$ is constant, and proceeds on that assumption without further analysis of the behavior of car $n - 1$ until the relative positions of n and $n - 1$ disagree 'substantially' from their expected values associated with constant $\dot{x}_{n-1}$. The level of 'substantial' may be an n-dependent parameter. When further decisions are indicated, either as just described or by a change in braking by $n - 1$ or $n - 2$, the evaluation of position and velocity is repeated.

The following are used for the estimated values of x_{n-1} and $\dot{x}_{n-1}$

$$\bar{x}_{n-1} = \text{estimated value of } x_{n-1} = x_{n-1} + RS\,|x_{n-1} - x_n| \tag{5}$$
$$\bar{\dot{x}}_{n-1} = \text{estimated value of } \dot{x}_{n-1} = \dot{x}_{n-1} + R'S\,|\dot{x}_{n-1} - \dot{x}_n|$$

where S = 'sensitivity', or driver observation accuracy and R,R' = pseudo-random numbers with mean zero, range ± 1, and rectangular distribution. Some justification for this evident simplification of the noise problem will be given in the section on parameters.

References p. 237/238

A 'substantial' error in the relative positions of cars $n - 1$ and n, sufficient to justify a new decision process, will be assumed to have occurred when

$$|x_{n-1}(t) - \bar{x}_{n-1}(t)| > KD \tag{6}$$

where $K =$ 'valor', a combination of sensitivity and courage, and

$$\bar{x}_{n-1}(t) = \bar{x}_{n-1}(t_D) + (t - t_D)\,\dot{\bar{x}}_{n-1}(t_D) \tag{7}$$

if the last decision was made at t_D.

We rewrite Eq. (1) to indicate these modifications

$$(\ddot{x}_n)_{t+\tau} = [C_1\,(\dot{\bar{x}}_{n-1} - \dot{x}_n) + C_2\,(\bar{x}_{n-1} - x_n - D) + C_3 B_{n-1} + C_4 B_{n-2}]_t \tag{8}$$

Each driver is characterized by a maximum acceleration $\ddot{x}_H$, a maximum deceleration $\ddot{x}_L$, a maximum velocity $\dot{x}_H$, and a minimum velocity zero. If the right side of Eq. (8) yields a value of acceleration which exceeds the acceleration limits or causes a velocity outside the imposed limits, the limiting value of acceleration is to be used instead.

The symbols introduced here are also listed in Table III.

THE COMPUTER PROGRAM

The simulation model has been programmed for an IBM 704 computer. The program consists of several subroutines representing:

1. The driver: as described.

2. The environment: The region traversed by the driver has a beginning and an end. Within the region there must be an opportunity to vary his behavior in the face of such obstacles as grades and curves. The driver follows another driver and so we generate a first driver, the platoon leader. The rate at which drivers arrive and are allowed to enter the region of interest is also within the sphere of our environmental control. Successive drivers may be programmed to exhibit differing parameter values.

3. The observer: is embodied in a printer and a cathode-ray tube exhibit.

THE FITTING OF PARAMETERS

In order to operate the simulation program it is necessary to assign explicit values to all the model parameters. We compare some aspects of our model to experimental car-flow results so as to deduce 'typical' parameter values for simulation use. We seek to make these values appropriate for traffic in tunnels.

References p. 237/238

Primary control parameters

Let us first consider the proposed control equation, Eq. (8), without driver error, without reactions to braking, and without limitations on acceleration or velocity. The equation then reduces to

$$\ddot{x}_n (t + \tau) = C_1 [\dot{x}_{n-1} (t) - \dot{x}_n(t)] + C_2 [x_{n-1}(t) - x_n(t) - D(t)] \tag{9}$$

where

$$D(t) = \alpha + \beta \dot{x}_n(t) + \gamma \ddot{x}_n(t)$$

We take the viewpoint that this is the primary control equation and that we can fit the parameters in Eq. (9) without reference to the others. Further, we suppose that the C_2 term is much less important than the C_1 term and so can be considered as a perturbation on

$$\ddot{x}_n (t + \tau) = C_1 [\dot{x}_{n-1} (t) - \dot{x}_n(t)] \tag{10}$$

We shall presently find support for the assumption that Eq. (10) represents the major part of Eq. (9). The other assumptions are based on both plausibility and necessity. We have neither the experimental data nor the enormous time required for fitting all the parameters simultaneously.

These assumptions permit us to first find C_1 and τ by fitting Eq. (10) to experimental data, and then to add the other term in Eq. (9) with data deduced independently.

CHANDLER, HERMAN, AND MONTROLL[1] have performed experiments in which the velocities, as functions of time, of a platoon leader and the car following him, were recorded as they proceeded for periods of 20 to 30 minutes on the General Motors test track in Detroit. The data was fitted to Eq. (10) with results as given in Table Ia. Note the very high correlation coefficients between theory and experiment.

FORBES, ZAGORSKI AND DETERLINE[4] also performed follow-the-leader experiments for the Port of New York Authority. They concentrated on recording leader–follower velocities and separations while the leaders performed short, violent maneuvers. Since they did not fit the results to Eq. (10), we obtained the raw data from the Port of New York Authority and performed our own analysis. Using runs of four different leader–follower combinations in the Lincoln Tunnel and two on an open highway, in each case we varied C_1 and τ to minimize the following quantity

$$\sum_{j=0}^{\text{end of run}} [\dot{x}_n^{(E)} (j\Delta t) - \dot{x}_n^{(\tau)} (j\Delta t)]^2 \tag{11}$$

where $\dot{x}_n^{(E)} =$ velocity of follower driver as found in experiment, $\dot{x}_n^{(\tau)} =$ velocity as given by solving Eq. (10), $\Delta t =$ time interval used to discretize data for ease in correlation. We used $\Delta t = 1/10$ second. The results are given in Table Ib.

TABLE I

EXPERIMENTALLY DETERMINED RELATIONSHIPS BETWEEN VELOCITY CONTROL PARAMETER C_1
AND REACTION DELAY TIME τ

The given pairs (C_1, τ) listed are the values best fitting Eq. (10) for a car immediately behind a platoon leader.

(a) Runs by CHANDLER *et al.*[1] Each run consists of 20 or more minutes of continuous operation on the General Motors test track.

Driver number	Reaction delay time τ in seconds	Velocity control parameter C_1 in seconds^{-1}	Correlation coefficient R between experimental data and Eq. (10) using the given coefficients
1	1.4	0.74	0.87
2	1.0	0.44	0.90
3	1.5	0.34	0.86
4	1.5	0.32	0.49
5	1.7	0.38	0.74
6	1.1	0.17	0.86
7	2.2	0.32	0.82
8	2.0	0.23	0.85

(b) Runs by FORBES *et al*[4]. Drivers 1 and 2 responded to sharp maneuvers on open highway in daylight. Drivers 3 to 6 responded to sharp maneuvers in level section of Lincoln Tunnel.

Driver No.	τ	C_1	R
1	1.0	0.7	0.86
2	0.5	1.3	0.96
3	0.6	0.8	0.91
4	0.5	1.0	0.87
5	0.7	1.1	0.96
6	0.5	1.0	0.86

We have plotted both sets of data in Fig. 1. Also plotted is the curve

$$C_1 = \frac{1}{2\tau} \tag{12}$$

It was noted in the first section of this paper that cars obeying Eq. (10) perform stably in reacting to a lead car's maneuvers if $C_1 < 1/2\tau$. The fact that many test drivers operate in the unstable region indicates that the equation does not present the complete control picture, despite the high correlations obtained with experiment.

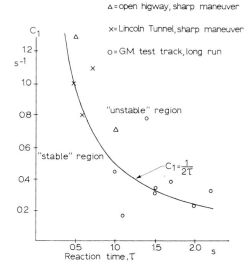

Fig. 1. Experimentally determined relationships between velocity control parameter C_1 and reaction delay time τ. The pairs (C_1, τ), as plotted, are the values best fitting $\ddot{x}_n (t + \tau) = C_1 [\dot{x}_{n-1} (t) - \dot{x}_n (t)]$ for the car immediately behind a platoon leader. Each point represents the behavior of a different driver. If $C_1 < (1/2\tau)$ in this equation, there is asymptotic stability, as $n \to \infty$, in response to a maneuver by the lead car, $n = 0$. By this we mean only that the velocity oscillations are finite and approach zero as $n \to \infty$, and not that there are no accidents. (see HERMAN et al.[10])

The graph also indicates that responses to sharp maneuvers are characterized by low values of τ and large values of C_1. The situation is reversed on long runs. Other observers support this result. The Traffic Engineering Handbook[9], page 81, notes that $\tau = 0.75$ s to 1.0 s is assumed by the American Association of State Highway Officials for the total perception–reaction–brake time in urban traffic, while $\tau = 2$ to 3 s is considered normal on the open highway.

In our tunnel simulations we generally assign

$$\tau = 0.75 \text{ s} \qquad C_1 = 0.5 \text{ (s)}^{-1} \qquad (13)$$

The value of τ is intermediate between the experimental values in tight maneuvering and those in open driving. We expect that normal tunnel traffic represents such an intermediate state. We have chosen a value for C_1 which is perhaps a little lower than Fig. 1 suggests for a 0.75 s reaction time. This was done because we wished to have the pair of values definitely on the stable side of $C_1 = 1/2\tau$.

Next, we consider the perturbation term in Eq. (9). We are now fortified in our belief that this C_2 term is small compared to the velocity control because, in using the velocity control alone, correlation coefficients of 0.80–0.95 were commonly obtained.

When a vehicle follows another at a constant velocity, so that the acceleration is zero, their separation is D. We have defined this as

$$D = \alpha + \beta \dot{x}_n + \gamma \ddot{x}_n$$

for the nth car. Now $\ddot{x} = 0$, and we find the equation fits very closely steady-state, velocity–headway data given on page 63 of the Traffic Engineering Handbook[9] when $\alpha = 15$ ft, $\beta = 0.98$ s. We also included the term $\gamma \ddot{x}_n$ in the equation for D because it was felt, that, when accelerating, a driver might choose a smaller head-way than when he is braking. Since we have no data on this matter, and since it is certainly very small, we set it equal to zero in all runs reported on here.

It seems that $\alpha = 15$ ft is rather small for long modern cars (the Traffic Engineering Handbook data is for traffic prior to 1950), and so we have taken

$$\alpha = 20 \text{ ft} \qquad \beta = 1 \text{ s} \qquad \gamma = 0 \tag{14}$$

for the simulation.

We use a somewhat indirect argument to find C_2. Suppose a car is travelling alone at its maximum speed $\dot{x}_H$. We assume that the car in front of it is so far away that the C_2 term of Eq. (9) contributes more to the desired acceleration than does the C_1 term. Then, the desired acceleration is positive, but, since the car is at its maximum velocity, the actual acceleration is zero. Now, suppose that the car overtakes another car whose velocity is zero. Eq. (9) now becomes

$$\ddot{x}_n (t + \tau) = -C_1 \dot{x}_n(t) + C_2 (x_{n-1} - x_n - D) \tag{15}$$

At the point where our driver first begins to react to the stationary vehicle, $\ddot{x}_n$ will change sign from positive to negative. Thus, the point where this occurs is characterized by

$$-C_1 \dot{x}_n + C_2 (x_{n-1} - x_n - D) = 0$$

or, rearranging terms and substituting for D

$$\frac{C_2}{C_1} = \frac{\dot{x}_n}{\Delta x_R - \alpha - \beta \dot{x}_n} \tag{16}$$

where Δx_R is the distance at which a driver going at $\dot{x}_n$ first decelerates when approaching a stationary obstacle.

Again, we do not have an experiment bearing directly on this situation. How-ever, the Highway Capacity Manual[7], page 38, has a graph which shows that cars on an open highway, at an average speed of 39.5 miles per hour, have velocities dependent to some extent on the velocities of the cars in front of them when their time headways are less than nine seconds. If this time headway can be taken as the correct value in the case of an approach to a stationary object, we have

$$\dot{x} = 58 \text{ ft/s} \qquad \alpha = 20 \text{ ft} \qquad \beta = 1 \text{ s} \qquad \Delta x_R = (58)\,(9) = 522 \text{ ft}$$

yielding

$$C_2/C_1 = 0.13\,\mathrm{s}^{-1}$$

from Eq. (16). Naturally, this value for the ratio is to be taken with strong reservations. In particular, the reaction distance Δx_R found above is for open highways where drivers do not encounter objects in front of them quite as frequently as in the urban tunnel environment. We feel that the value prevailing in the tunnel may be much larger and so have used

$$C_2/C_1 = 0.25\,\mathrm{s}^{-1} \quad \text{or} \quad C_2 = 0.125\,\mathrm{s}^{-2} \tag{17}$$

in the tunnel bottleneck simulation. This larger value is justified by the very strong follow-the-leader psychology of congested rush hour traffic, in which most drivers will attempt to rapidly reach a minimum safe headway.

Braking-reaction control parameters

No experimental data is available at the time of writing this paper on suitable values for C_3 and C_4 of Eq. (8). It was hoped that some results would be obtained from instrumented vehicle runs by HERMAN et al. at General Motors. Since these are not yet available, we have set

$$C_3 = C_4 = 0 \tag{18}$$

in the simulation. However, we can have cars react to braking by re-assessing their position as described in the first section of this paper. For this we need only to evaluate F. We asserted that a car is braking, rather than coasting, when

$$\ddot{x}_n < F\dot{x}_n$$

The Traffic Engineering Handbook[9] has some data on this in Table 61, page 70. The relationship is not as linear as implied by our formula. However, the formula is a fair approximation when

$$F = 0.04\ (\mathrm{s}^{-1}) \tag{19}$$

Acceleration and velocity limits

All data used here is from the Traffic Engineering Handbook[9], pages 58–68. The absolute maximum acceleration of passenger cars is on the order of 15 ft/s² in low gear and 6.5 ft/s² in high. However, the average accelerations, for vehicles accelerating from a standstill to normal highway speed, are much lower. These range from 4.6 ft/s², at 10 miles per hour, to 3.8 ft/s², at 30 miles per hour, with still lower values at higher speeds. It seems to us that these values, for acceleration from rest, probably represent maximum values for vehicles varying their velocities long after they have started moving. In our effort at linearization, we select 4 ft/s², independent of velocity, as a typical maximum. Trucks will naturally have a considerably lower capability.

The maximum braking acceleration is about -30 ft/s². A value of -22 ft/s² is sufficient to throw unprepared passengers out of their seats. It is stated that about -20 ft/s² is a practical maximum.

The maximum velocities drivers assume are dependent largely on external circumstances and speed limits. We have used rather low values in our simulation because the tunnel environment does not encourage speeds above 40 miles per hour even when there is enough room.

Generally, we have set

$$\ddot{x}_H = \begin{cases} 4 \text{ ft/s}^2 \text{ for cars} \\ 2 \text{ ft/s}^2 \text{ for trucks} \end{cases} \qquad \begin{aligned} \ddot{x}_L &= -20 \text{ ft/s}^2 \\ \dot{x}_H &= 44 \text{ to } 60 \text{ ft/s} \end{aligned} \qquad (20)$$

Driver error parameters

In the first section of this paper it was postulated that the driver of car n would err in estimating the distance and velocity of car $n - 1$ as follows

$$\tilde{x}_{n-1} = x_{n-1} + RS \, |x_{n-1} - x_n|$$
$$\tilde{\dot{x}}_{n-1} = \dot{x}_{n-1} + R'S \, |\dot{x}_{n-1} - \dot{x}_n|$$

Here $\tilde{x}_{n-1}$, $\tilde{\dot{x}}_{n-1}$ are the estimated and, hence, perhaps erroneous values, R, R' are random numbers from a rectangular distribution with mean zero and range ± 1 and S is the sensitivity, or accuracy of driver observation. We admit at once that this formulation is almost frivolous because our ignorance of the actual error mechanism is profound. However, we do require driver errors because drivers assess headway and velocity with the limited precision provided by their variably effective instruments, eyes, and brains. Under the circumstances, it is no great crime to use a rectangular random number distribution. Further, it is reasonable to suppose that the average error in judging headway will increase as that distance itself increases. Similarly, the error in judging relative velocity should be trivial if this quantity is zero and should increase with the velocity difference.

To reduce the number of decisions a driver makes from a value of one per integration interval to something more reasonable, we have adopted a procedure where car n will assume that $n - 1$ goes at a constant speed unless the latter's position becomes so far removed from that postulated by this assumption that n will be roused to a new decision. As given by Eq. (6) the condition for this is

$$|x_{n-1}(t) - \tilde{x}_{n-1}(t)| > KD \quad \text{where} \quad \tilde{x}_{n-1}(t) = \tilde{x}_{n-1}(0) + t\tilde{\dot{x}}_{n-1}(0)$$

If the 'valor' parameter K is large, the driver is sleepy, foolhardy or brave, and makes decisions rarely.

We evaluate S and K by operating the model drivers with all other parameters as defined up to this point (listed under 'typical' car in Table III) and compare their behavior with that of real drivers. Let a model platoon leader go over a course at constant speed. Let him be followed by several platoon members, all with the same values of S and K. Since the platoon leader does not change his speed, the average acceleration of the followers over the whole trajectory is zero. This average acceleration $\langle \ddot{x}_D \rangle$ is

$$\langle \ddot{x}_D \rangle = \frac{1}{n-1} \sum_{j=2}^{n} \left(\frac{1}{T} \int_0^T \ddot{x}_j dt \right) = 0 \tag{21}$$

Here, $j = 2$ corresponds to the first car behind the platoon leader of an n-car platoon moving over a course of such length that it takes a time T for it to be traversed. We use the subscript D on $\langle \ddot{x}_D \rangle$ because random acceleration is often referred to as acceleration 'dispersion'. The variance of the drivers' acceleration gives a measure of the random, or noise, acceleration

$$\sigma_D{}^2 = \langle \ddot{x}_D{}^2 \rangle - \langle \ddot{x}_D \rangle^2 = \langle \ddot{x}_D{}^2 \rangle = \frac{1}{n-1} \sum_{j=2}^{n} \left(\frac{1}{T} \int_0^T \ddot{x}_j{}^2 \, dt \right) \tag{22}$$

We call σ_D the acceleration dispersion. The computer program has been instrumented to find this quantity. Table II gives σ_D as obtained by simulation for various values of S and K. We have tabulated only cases where $K \geq S$ because $K < S$ implies that a driver will make a decision on information of greater accuracy than he is capable of obtaining. In seeking suitable model values for S and K, we set $K = 2S$ as a reasonable guess on arbitrary grounds.

HERMAN, MONTROLL, POTTS AND ROTHERY[10] have experimented to find σ_D under typical driving conditions. Their results, on a preliminary basis, are:

(a) $\sigma_D = 0.01\, g = 0.32$ ft/s^2 for a car proceeding alone on a highway at 35 miles per hour.

(b) $\sigma_D = 0.03\, g = 0.96$ ft/s^2 for a car proceeding smoothly in moderate traffic at 35 miles per hour.

(c) $\sigma_D = 0.06\, g = 1.92$ ft/s^2 for a car trying to go at 45 miles per hour in traffic averaging 35 miles per hour. Passing was permitted.

Since our computer experiment was conducted at 48 ft/s, or 32.6 miles per hour it corresponds well to HERMAN's situation (b). Setting $K = 2S$, Table II shows that

$$S = 0.125 \qquad K = 0.250 \tag{23}$$

yielding $\sigma_D = 0.87$ ft/s^2, gives good agreement with (b) above. Consequently, these were selected for the 'typical' driver.

References p. 237/238

Résumé: The parameters selected for the 'typical' driver, as given by Eqs. (13), (14), (17), (18), (19), (20) and (23), are summarized in Table III.

TABLE II

ACCELERATION NOISE

S = sensitivity, defined by Eq. (5)
K = valor, defined by Eq. (6)
σ_D = standard deviation of acceleration, defined by Eq. (22)

S	K	σ_D [ft/s^2]
0	0	0
0	0.0625	0 *
0.0625	0.0625	0.28
0.0625	0.125	0.41
0.125	0.125	0.56
0.125	0.25	0.87
0.25	0.25	1.13
0.25	0.5	1.71

* As long as S equals zero, no accelerations occur because the platoons start their motion with inter-vehicle separations at the desired headway value. With perfect observation, these headways are maintained throughout and no acceleration is called for.

TABLE III

DRIVER PARAMETERS

Symbol	Description	Units	'Typical' car value	Truck value	
				Normal	When affected by bottleneck of 2nd kind
C_1	velocity control	[s^{-1}]	0.5	0.5	
C_2	headway control	[s^{-2}]	0.125	0.125	
C_3	brake factor on car $n-1$	[ft.s^{-2}]	0	0	
C_4	brake factor on car $n-2$	[ft.s^{-2}]	0	0	
τ	reaction time	[s]	0.75	0.75	
$\ddot{x}_H$	maximum acceleration	[ft.s^{-2}]	4.0	2.0	1.0
$\ddot{x}_L$	minimum acceleration	[ft.s^{-2}]	−20.0	−20.0	
$\dot{x}_H$	maximum velocity	[ft.s^{-1}]	51.0	51.0	
D	desired headway $D = \alpha + \beta\dot{x} + \gamma\ddot{x}$	[ft]			
α_c	collision headway	[ft]	16.0	30.0	
α		[ft]	20.0	40.0	
β		[s^{-1}]	1.0	1.0	
γ		[s^{-2}]	0	0	
F	braking parameter	[s^{-1}]	0.04	0.04	
K	'valor', decision parameter	—	0.25	0.25	
S	'sensitivity', noise parameter	—	0.125	0.125	

Unless stated otherwise, bottleneck values are the same as nonbottleneck values.

Flow Bottlenecks

INTRODUCTION: TWO KINDS OF BOTTLENECKS

A bottleneck is a stretch of roadway with a flow capacity less than that of the road ahead, called downstream, or behind, called upstream. We consider bottlenecks as found in tunnels, although there is no intent to claim that these differ much from those found elsewhere.

Suppose that the roadway, including the bottleneck, is empty of vehicles except for an isolated compact platoon of cars. We label the cars according to their position j relative to the platoon leader. The leader is car zero, and the last car in the platoon is car n, so $j = 0, 1, 2, \ldots, n$. Let $\Delta^{(j)}$ be the time headway of the jth car in a platoon. The time headway is the time interval, as observed at a stationary point, between the passage of the front end of the $(j - 1)$th car and that of the jth car past the observation point. If a platoon flow experiment is repeated a number of times, it is possible to obtain an average, $\bar{\Delta}^{(j)}$, of this quantity.

Within the platoon, the cars are 'following the leader' so that their time headways correspond to the maximum practical flow. While the entire platoon is upstream from the bottleneck, these time headways correspond to the maximum flow possible in this non-bottleneck region. When the platoon enters the bottleneck, the time headways will increase because the flow capacity is now less.

If a number of such isolated platoons are observed, one can find the average time headways $\bar{\Delta}^{(j)}$ for the jth car as a function of position both within and outside the bottleneck. While the platoon is upstream from the bottleneck, we do not expect $\bar{\Delta}^{(j)}$ to vary much with j. When the bottleneck is entered $\bar{\Delta}^{(j)}$ will increase. We can classify bottlenecks into two groups on the basis of whether this increase varies with the vehicle position j within the bottleneck. If the increase is the same for all values of j, we define the bottleneck as one of the first kind. If it varies with j, we have a bottleneck of the second kind. We clarify this classification with some examples.

In a bottleneck of the first kind, a driver requires a longer time headway than outside the bottleneck, but this headway does not depend on his position in the platoon. The bottleneck time headway will be the same for a member of a small platoon as for any member of an infinitely long platoon (corresponding to the continuous flow situation). This condition will be met wherever a driver chooses his bottleneck headway independently of the behavior of the vehicles in front of him. It is practical for him to do so only when the behavior of the vehicles in front is fairly smooth, without violent acceleration, and when the vehicles have the capability of assuming the desired headway quickly. The latter condition may not be met on a hill, for example, because there some drivers may not be able to

References p. 237/238

either maintain speed or accelerate to the velocity of their leaders. Thus, these drivers will be unable to achieve their desired headways. Bottlenecks which conform to our requirements probably include gradual curves, where drivers increase their headways because of reduced visibility, and sections of inferior roadway, clearly signalized in advance, where drivers slow down to avoid damage to their vehicles. In the next section we present a simulation of a bottleneck of the first kind.

A bottleneck of the second kind is one in which a driver will, on the average, react differently if he is far back in a platoon than if he is near the platoon leader. His behavior is thus much dependent on the bottleneck behavior of the vehicles in front of him. For example, the average time headway may increase with the distance from the platoon leader as a result of limitations on vehicle performance. We shall see, from an upgrade bottleneck example, that a low bottleneck acceleration capability will result in this behavior. This is because flow through the bottleneck decreases with decreasing velocity and because a velocity decrease by any vehicle in the bottleneck cannot be rapidly rectified by its followers. Rather, they will tend to amplify the fall-off in performance.

Obviously, the best way to alleviate a bottleneck is to remove its cause. When this is not practical, special traffic controls or driver education may help. We have found that one very simple type of control, namely the deliberate platooning of vehicles, may have a very beneficial effect on flow through bottlenecks of the second kind, as shown later in this paper.

It will be noted that vehicle flows in the simulation are extremely high when there are no bottlenecks. For example, in one case we obtained a flow of almost 2200 cars per hour. In actual traffic, a flow of this size is found only on some multilane highways under unusual circumstances. Our high figure is not really suprising. Such a flow is the ideal maximum attainable only if there is no bottleneck whatever, if all drivers are reasonably alert, and if all try to go as fast as possible.

A BOTTLENECK OF THE FIRST KIND: GRADUAL LOCALIZED VARIATION IN DESIRED HEADWAY

Description

Suppose we have a tunnel with a curve in it sufficiently sharp to obscure substantially the visibility of the road for the driver negotiating it. Under these circumstances, the driver tends to increase his headway so that he may have more time to cope with unforeseen emergencies.

If he reacts very sharply to the beginning of the curve, his braking will produce repercussions behind him, and the bottleneck will not be one of the first kind. For then the followers will brake at a more rapid rate than that at which they are capable of accelerating. If an isolated platoon passes through, this sudden activity will not stabilize until a number of cars have passed through the bottleneck. During the initial period, the average time headway of the jth car in the platoon will increase sharply with j, as in the second kind of bottleneck. We believe that a true bottleneck of the first kind is rare. Its existence requires that the driver adjusts himself gently to the new situation so that his followers adopt an efficient waiting attitude, as required, and then follow without unnecessary lost time.

Simulation

We define a bottleneck as a 1024 ft section of roadway, starting 2048 ft from the tunnel entrance. Through this, we feed 'typical' cars as defined in Table III.

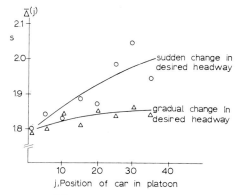

Fig. 2. Average time headway, $\bar{\Delta}^{(j)}$ as a function of position within platoon, in a bottleneck caused by an increase of desired space headway. The circles represent averages as found in simulation.

When not in the bottleneck, the cars try to follow each other as per the table. In the bottleneck we assume $\beta = 1.25$ s. To achieve a bottleneck of the first kind we allow the driver 256 ft in which to make the change gradually. Thus

$$\beta = \begin{cases} 1 \text{ s} & x \leq 2048 \text{ ft and } x \geq 3072 \text{ ft} \\ \left(1 + \dfrac{x - 2048}{1024}\right) \text{ s} & 2048 \text{ ft } \leq x \leq 2304 \text{ ft} \\ 1.25 \text{ s} & 2304 \text{ ft } \leq x < 3072 \text{ ft} \end{cases}$$

In contrast, we also try runs where β changes its value abruptly at 2048 ft.

Several runs were made with both schemes. For each, the average time headway

$\bar{J}^{(j)}$ of the jth car in an isolated platoon is plotted as a function of j in Fig. 2. It is evident from this graph that the sudden change in β causes a bottleneck of the second kind. However, with a gradual change, we obtain the expected behavior.

A BOTTLENECK OF THE SECOND KIND: LOCALIZED ACCELERATION LIMIT

Description

A typical underwater tunnel has a downgrade followed by a level stretch and then by an upgrade to the surface. Such a tunnel often has its limiting bottleneck on the upgrade, particularly at the point where the upgrade begins.

We advance an obvious explanation. All the vehicles have a reduced acceleration capability on any upgrade. Some, trucks for example, have a substantially reduced capability. Not all vehicles in the stream will have the same velocity at any given tunnel point. There will be a fluctuation about the mean velocity. Consider now the foot of the upgrade. As traffic passes, there will eventually appear at this point a vehicle, at a speed much below average, which has a poor upgrade acceleration capability. This vehicle will then remain at a relatively slow speed for some time, even if it attempts an increase. If followed by many vehicles with good acceleration, the flow is almost unimpeded, since these followers will gradually restore the normal velocity average. But if one of the near followers also has a poor acceleration on the upgrade, this follower, when forced to slow down, will also recover slowly. Thus, if the proportion of poor accelerators is high, the average speed of vehicles will eventually stabilize at a value much below that which would prevail in the absence of a bottleneck. At this lower speed, the flow will generally be less.

Simulation

Method. We postulate two types of vehicles, to be called cars and trucks. The trucks are impeded by an upgrade, unlike cars, and are longer. The parameters selected to describe the vehicles are intended to be realistic for passenger cars and for large trucks. Even normally, a truck has only half the car's acceleration capability of 4 ft/s². Within the upgrade bottleneck, this is reduced further to 1 ft/s² or less. The response characteristics of both types of vehicles are shown in Table III. There, cars are called 'typical'. The simulation tunnel is 4096 ft long. The upgrade bottleneck begins at 2048 ft from the entrance and is 1024 ft long. In all runs the model platoon leaders go at thirty miles per hour (44 ft/s) so that the results are comparable. In life, of course, the platoon leaders' velocities differ and, as we shall see later, this variation must be accounted for when planning improvement schemes based on platooning.

References p. 237/238

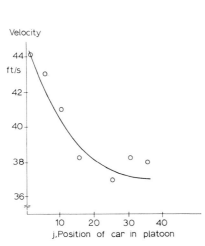

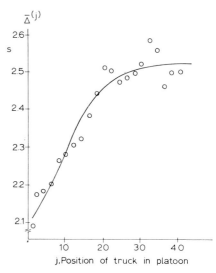

Fig. 3. Bottleneck velocity as a function of vehicle position in platoon. Results for trucks in acceleration-limit simulation, observed 256 ft downstream from start of bottleneck. The points present averages of observations. The curve has been fitted to these and has no mathematical significance.

Fig. 4. Average time headway, $\bar{\Lambda}^{(j)}$, as a function of platoon position in a bottleneck of the second kind. Results for trucks, with acceleration capability cut in half within a bottleneck 1024 ft long. Observation is at end of bottleneck. The points present averages of observations. The curve has been fitted to these and has no mathematical significance.

The results below are statistical in nature, since the drivers' responses vary (neither S, driver sensitivity, nor K, the decision parameter, equals zero). In consequence, each experiment consisted of several runs under the same conditions.

Results for platoons composed only of trucks. The simulation was run first with platoons composed entirely of the trucks described above. In Fig. 3, we plot the average velocity of the jth vehicle in the platoon at 2304 ft from the tunnel entrance or 256 ft downstream from the start of the bottleneck. As anticipated, this decreases as j, the numerical position of a vehicle relative to the platoon leader, increases. While the platoon leader passed through the bottleneck at an average speed of 44 ft/s, the steady state velocity, reached at about the 25th vehicle in the platoon, is only 37 ft/s. Because of the poor bottleneck acceleration, there is little chance that the later vehicles in the platoon will ever go faster. Once several vehicles have passed through at a low speed and the flow has decreased, the later vehicles of a long platoon will have to reduce their speed even before they reach the bottleneck. This is because the platoon flow is greater upstream before the

References p. 237/238

bottleneck than in it. Since the situation of unequal flow cannot prevail long, the flow upstream will become equal to that in the bottleneck as the platoon length n approaches infinity.

In the introductory section of this part of the paper, we defined the average time headway of the jth vehicle in a platoon, $\bar{\Delta}^{(j)}$. It should be noted that the corresponding instantaneous flow is

$$\bar{q}^{(j)} = \frac{3600}{\bar{\Delta}^{(j)}}$$

vehicles per hour if $\bar{\Delta}^{(j)}$ is in seconds.

The average time headways found at 3072 ft, the end of the bottleneck, as a function of j, are plotted in Fig. 4. These time headways, as shown, are the truly significant quantities emerging from the simulation. The velocities, given previously, were of interest, but did not in themselves indicate the noteworthy deterioration in flow with increasing platoon length implied by Fig. 4. Here, the time headways of the first few trucks behind the leader, at about 2.15 s each, correspond to a flow of 1650 trucks per hour. The time headways of trucks 25 or more places behind, at 2.5 s, correspond to a flow of 1425 trucks per hour. This corresponds to the steady-state flow.

We shall find it convenient to use the time headways of Fig. 4 in a somewhat different manner. The time taken by an n-car platoon in passing a given point in the tunnel is

$$\overline{\sum_{j=1}^{n} \Delta^{(j)}} = \sum_{j=1}^{n} \bar{\Delta}^{(j)}$$

In Fig. 5 we plot

$$H_n = \frac{1}{n} \overline{\sum_{j=1}^{n} \Delta^{(j)}}$$

the average time headway in a platoon of length n, based on the same data as used in Fig. 4. However, now $1/H_n$ is the flow level achieved by the n-car platoon as a whole, a most useful quantity. We note that the relations of Fig. 5 are represented almost exactly by a curve of the form

$$H_n = A - Be^{-Cn} \tag{23}$$

where

$$A = 2.53 \text{ s} \qquad B = 0.44 \text{ s} \qquad C = 0.0347$$

This curve was obtained without any theoretical basis and merely represents a numerical fit. However, its excellent correspondence to the data suggests that a similar exponential decay formula would be most welcome in a theoretical model. Unfortunately, the model we have been able to develop, presented later in this section, does not yield this relation, although the behavior is similar.

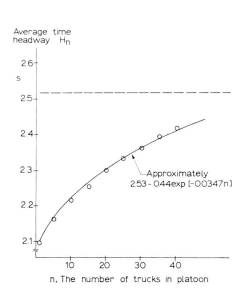

Fig. 5. Average time headway in a platoon of length n in a bottleneck of the second kind. The simulation is the same as presented in Fig. 4. The time length of the platoon of length n is n times the average presented here.

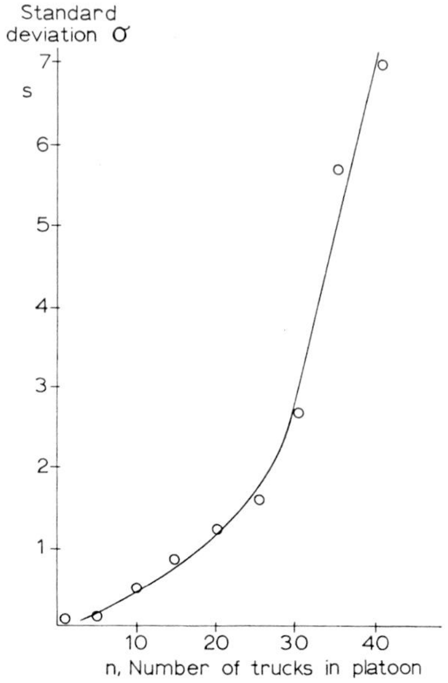

Fig. 6. Standard deviation of the time length of a platoon in a bottleneck of the second kind.

Note that the fitted curve shows clearly that the average time headway approaches a constant value independent of n as n approaches infinity.

The time length of a platoon varies greatly between repeated runs. In Fig. 6 we plot, as a function of n, its standard deviation

$$\sigma_{\text{time length, } n} = \left\{ \sum_{k=1}^{s} \left(\sum_{j=1}^{n} \varDelta^{(j)} \right)^2 - \left(\sum_{k=1}^{s} \overline{\sum_{j=1}^{n} \varDelta^{(j)}} \right)^2 \right\}^{\frac{1}{2}} s^{-\frac{1}{2}}$$

where $s = 4$ is the number of observations employed. It is seen that, as the number of vehicles in the platoon approaches infinity, the standard deviation in overall platoon time length approaches approximately $(0.43\,n - 10)$ seconds. However, the variation increases more slowly for small n because, as we have seen, small platoons succeed in passing the bottleneck with little change in their disciplined regular follow-the-leader spacing and velocity. It is likely that the variations in

References p. 237/238

platoon length found in actual traffic will be larger than found here because of variations between individuals not included in the model representation.

The bottleneck flow capacity varies with the maximum acceleration available to the trucks within it. In Fig. 7 we show $\bar{\Delta}^{(j)}$ as a function of j and of the maximum bottleneck acceleration. The dependence is clear, even though only three values of acceleration are plotted. Note that, as might be expected, a value

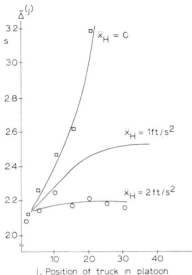

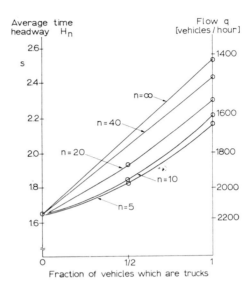

Fig. 7. Average time headway $\bar{\Delta}^{(j)}$, as a function of acceleration capability and position within platoon in a bottleneck of the second kind. The bottleneck is the same as in Fig. 4, whose curve is presented again here. The maximum acceleration is 2 ft/s² outside the bottleneck.

Fig. 8. Comparison of average time headways in a bottleneck of the second kind, as a function of platoon length and of the fraction of trucks; n is the number of vehicles in the platoon.

$\ddot{x}_H = 0$ yields values for $\bar{\Delta}^{(j)}$ which increase without bound as j approaches infinity. Thus, the flow will be zero unless the number of vehicles in a platoon is kept small. There is a simple explanation for this situation. When $\ddot{x}_H = 0$ in the bottleneck, a slow vehicle will traverse it without increasing its speed. Then, its followers will be forced to adopt its slow pace as a maximum. Eventually, one of them will drop to a lower speed, at least momentarily, and will be unable to recover. Ultimately, a vehicle will stop and all flow will cease.

Results for platoons composed of both cars and trucks. We next generate platoons containing random mixtures of 'typical' cars and of trucks as defined in Table III.

References p. 237/238

The trucks have their maximum acceleration reduced from 2 to 1 ft/s² in the bottle-neck. The cars are shorter than the trucks and will therefore try to maintain a shorter headway.

Cars alone are unaware of the bottleneck and so their flow is independent of platoon length. When there is a mixture of cars and trucks, the cars, having superior acceleration, follow the bottleneck-affected trucks closely, thus forming sub-platoons led by trucks and composed of cars. Thus, the cars do not contribute to the bottleneck slowdown and the situation with both cars and trucks in the stream is intermediate between that for cars alone and that for trucks alone. In Fig. 8, we plot the average time headways of vehicles in a platoon of n vehicles as a function of n and of the fraction of trucks in the car–truck mix. We show the corresponding flow q in vehicles per hour.

Alleviation: proposed experimental procedure for establishing control parameters

The fact that the time headway of the jth vehicle in a platoon increases with j leads us to consider enhancing flow by deliberate platooning. The introduced gaps must be minimized to maximize the flow. Thus one would seek to have them as short and as few as possible. But few gaps correspond to long platoons. As the platoons grow longer the platooning flow advantage decreases. Thus, we face an optimizing problem, namely, the determination of gap length and platoon size for maximum flow. A location–time plot of the trajectories of the vehicles in two successive platoons will clarify the nature of the variables to be considered. Fig. 9 is such a diagram which has been idealized for clarity. The platoon leaders will not all go at the same velocity, and so there will be a spread of values about t_{1e}, the time taken by the platoon leader to go from the tunnel entrance control point to the end of the bottleneck. The variation in the value of T_{ne}, the time taken by an n-car platoon to pass the end of the bottleneck, has already been noted. The standard deviation, $\sigma(T_{ne})$, of this time interval was shown in Fig. 6 for one all-truck simulation. Similarly, there is a variation in T_{ns}, the time length of the platoon at the entrance control point. Suppose that the deliberately introduced gap at the control point, δ_{ns} is made constant. If we had available the variances of the three distributions for t_{1e}, T_{ne} and T_{ns}, we could estimate this for δ_{ne}, the gap between platoons at the end of the bottleneck

$$\sigma^2(\delta_{ne}) \cong \sigma^2(t_{1e}) + \sigma^2(T_{ne}) + \sigma^2(T_{ns})$$

This relationship would be an equality if there were not interdependence between the three variables. It is probable that this is not the case.

If $\sigma(\delta_{ne})$ were zero, we could make $\delta_{ne} = 0$ and achieve an easy maximum flow. But it is definitely nonzero and so we must introduce an average end gap, $\bar{\delta}_{ne} > 0$,

References p. 237/238

to prevent the platoons from running together before the end of the bottleneck. If two platoons do merge at any substantial distance before the bottleneck ends,

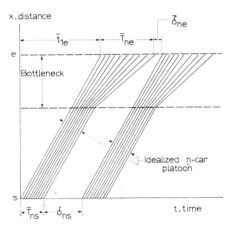

Fig. 9. Platooning to increase flow at bottleneck of the second kind.

$\bar{t}_{1e}$: Average time taken by the platoon leader to go from the control point (at tunnel entrance) to the end of the bottleneck.

$\overline{T}_{ns}$: Average time taken by n-car platoon to pass the control point.

$\overline{T}_{ne}$: Average time taken by the platoon to pass the end of the bottleneck.

δ_{ns}: Deliberate platooning gap at the control point.

δ_{ne}: Average platooning gap at the end of the bottleneck.

it is unlikely that the desirable gapping process will again be established unless a special extra-long control point gap is introduced. This is so because the steady-state continuous flow capacity of the bottleneck is less than the flow achieved with the aid of the platooning mechanism.

There are two ways to avoid this merger catastrophe. One is to make no normal provision for it in the entrance gap but, instead, to post an observer at the bottleneck to notify the control point when the platoons are merging too early so that the gaps may be temporarily increased in size. The second is to adjust the control gap so that the average value of δ_{ne} is large compared to $\sigma(\delta_{ne})$.

We do not have experimental data to illustrate the problem completely by simulation. In particular, we do not know how the platoon leader will proceed, both when isolated and when approaching the tail of the preceding platoon. However, we can make a limited illustration of the flow improvement possible for our simulation bottleneck. Consider the first example, namely, trucks whose maximum acceleration is halved in a 1024-ft bottleneck. Rather than arbitrarily predict a correct $\bar{\delta}_{ne}$, we compare several values.

References p. 237/238

From Eq. (23) we have approximately

$$\bar{T}_{ne} \equiv nH_n = n\,\{2.53 - 0.44\exp\,[-0.0347\,n]\}\ \text{seconds} \qquad (24)$$

$$\bar{T}_{ns} = 2.09\,n\ \text{seconds}$$

The flow is

$$q = \frac{n}{\bar{T}_{ne} + \delta_n} = 3600\left\{2.53 - 0.44\exp\,[-0.0347n] + \frac{\bar{\delta}_{ne}}{n}\right\}^{-1}\ \text{cars/hr}$$

provided that the platoons do not merge. This flow is shown as a function of n for various values of $\bar{\delta}_{ne}$ in Fig. 10.

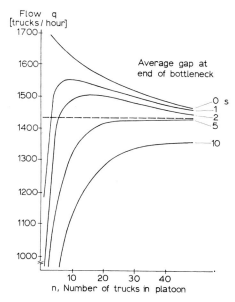

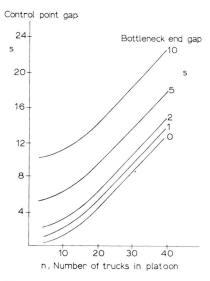

Fig. 10. Effect on flow of gap size at end of bottleneck. This example is derived from the simulation of trucks passing through a bottleneck of the second kind, as shown in Fig. 5.

Fig. 11. Relation between gap size at end of bottleneck and gap size at control point as a function of platoon length. This example is derived from the simulation of trucks passing through a bottleneck of the second kind, as shown in Fig. 5.

It is seen that $\bar{\delta}_{ne}$ must be less than 5 s for any improvement in flow. A value of 2 s gives a maximum increase of 75 cars per hour for platoons 17 cars long. The case $\bar{\delta}_{ne} = 0$ is plotted only as a limit, since then there would not really be any platooning.

References p. 237/238

We relate $\bar{\delta}_{ne}$ to δ_{ns}

$$\bar{T}_{ne} + \bar{\delta}_{ne} = \bar{T}_{ns} + \delta_{ns}$$

From Eq. (24) then follows

$$\delta_{ns} = 0.44\, n\, \{\mathrm{1} - \exp\left[-0.0347\, n\right]\} + \bar{\delta}_{ne}\ \text{seconds}$$

This is plotted as a function of n and of $\bar{\delta}_{ne}$ in Fig. 11. The quoted example of a 2-second end gap for a 17-car platoon gives a control gap of 5.5 s. To this must be added about two seconds, the time headway of the platoon leader. The total 7.5 s is the time interval to be introduced between platoons.

Since reasonably large gaps result in small or non-existent flow improvements, we suspect that platooning, without flexible gap timing, will work only for bottle-necks of the second kind much more severe than the one simulated here. On the whole, it seems most worth while to investigate practical procedure for controlling gap lengths on the basis of immediate performance rather than making them constant without regard to minute by minute performance.

To determine whether a particular real bottleneck is susceptible to alleviation by platooning, some experiments must be performed. First, we need to know the average time headway and its variation as a function of position in an isolated platoon both at the end of the bottleneck and at the proposed control point. This information can be obtained when the roadway is not congested because enough members of the public to form a large compact platoon may be accumulated at the control point before allowing the platoon to proceed. As in our simulation, the time headways of all platoon members should be observed both at the control point and at the end of the bottleneck. In addition, the time taken by the platoon leader to reach the end of the bottleneck should be observed. Enough long platoons should be run to obtain good statistics. There is little need to test platoons of varying length since the behavior of a short platoon should differ only slightly from that of an equivalent sub-platoon headed by the long-platoon leader.

The information gleaned from the described experiment is enough to tell whether platooning is worth while and to provide initial estimates of platoon and gap sizes. If the congested traffic is to be controlled with a variable gap size, it would now be possible to implement the solution, making small adjustments as needed.

However, if the gap size is to be constant, as preferable with automatic control signalling or little manpower, the situation must be studied further. We have to find out how successive platoons merge and what time interval constitutes a gap between a platoon leader and the tail of the preceding platoon. We can still test during non-rush hour periods by accumulating two platoons' worth of cars for each experiment. Then, we repeat the previous experiment, but with both platoons seperated by an input gap which is varied until a satisfactory end-of-bottleneck condition is determined.

Theoretical model of a bottleneck of the second kind

We present a simple theoretical model for the acceleration-limit bottleneck. At its present state of development, the model does not represent the process very well in a quantitative sense. However, it is conceptually instructive and may well lead to a more sophisticated formulation.

Let vehicles approach the bottleneck with a velocity distribution over a small range about the average value. Let this distribution be approximated by a discrete rectangular distribution, *i.e.* the probability distribution

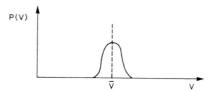

Fig. 12a. Velocity distribution over a small range about the average value.

replaced by

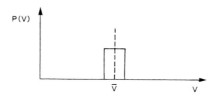

Fig. 12b. Rectangular velocity distribution.

replaced by

$$V_1 \text{ with a probability } P_1^{(0)} = 1/r$$
$$V_2 \text{ with a probability } P_2^{(0)} = 1/r$$

$$V_r \text{ with a probability } P_r^{(0)} = 1/r$$

where $(V_j - V_{j-1}) = (V_r - V_1)/(r - 1)$, $\qquad j = 2, 3, \ldots, r$.

The superscript (o) distinguishes this distribution from others defined later. The distribution can be represented by the probability vector

$$\mathbf{P}^{(0)} = (P_1^{(0)}, P_2^{(0)}, \ldots, P_r^{(0)}) = (1/r, 1/r, \ldots, 1/r) \tag{25}$$

We postulate a simple bottleneck mechanism. If car n enters the bottleneck at a speed V_j, car $(n + 1)$ must change its speed to V_j if it was greater than V_j.

Otherwise, he continues at the same speed as before. This is reasonable: Acceleration is slow in the bottleneck and so vehicles must slow down to about the speed of those ahead, since the latter cannot rapidly increase their speed. Clearly, as the number of vehicles in a platoon approaches infinity, the bottleneck velocity distribution will approach $P(V_1) = 1$.

The problem: What is the velocity distribution in the bottleneck as a function of n, the index number for cars in the platoon?

For the platoon leaders, $n = 1$, the velocity distribution is

$$P^{(1)} = P^{(0)} = (1/r, \, 1/r, \, \ldots, \, 1/r)$$

since these cars are not delayed.

Given the distribution $P^{(n)}(V)$, we can formulate $P^{(n+1)}(V)$. For the discrete rectangular arrival distribution and the formulated bottleneck rule, the relation is

$$P^{(n+1)} = P^{(n)} \cdot T \tag{26}$$

where

$$T = \begin{bmatrix}
1 & 0 & 0 & 0 & \cdot & \cdot & & \cdot & 0 \\
\dfrac{1}{r} & \dfrac{r-1}{r} & 0 & 0 & \cdot & \cdot & & \cdot & 0 \\
\dfrac{1}{r} & \dfrac{1}{r} & \dfrac{r-2}{r} & 0 & \cdot & \cdot & & \cdot & 0 \\
\cdot & \cdot & \cdot & \cdot & & & & & \cdot \\
\cdot & \cdot & \cdot & \cdot & & & & & \cdot \\
\dfrac{1}{r} & \dfrac{1}{r} & \dfrac{1}{r} & \dfrac{1}{r} & \cdot & \cdot & & \dfrac{1}{r} &
\end{bmatrix} \tag{27}$$

The algebraic resemblance to a Markov process is complete. Rows of the transition matrix correspond to velocity states of car n. Columns correspond to states of car $(n + 1)$.

Rather than use the general algebraic approach to the solution of finite Markov chains[3] we proceed in an elementary manner, developing the general solution by mathematical induction. We exhibit the solution

$$P_j^{(n)} = \frac{(1 + r - j)^n - (r - j)^n}{r^n} \tag{28}$$

To prove this is correct, we show
(1) $P_j^{(1)}$ is correct, and
(2) if $P_j^{(n)}$ is correct, then $P^{(n+1)}$ is correct.

Proof:

(1)
$$P_j^{(1)} = \frac{(1 + r - j) - (r - j)}{r} = \frac{1}{r}$$

This is the defined initial velocity distribution and, hence, correct.

(2) We have

$$\boldsymbol{P}^{(n+1)} = \boldsymbol{P}^{(n)} \cdot \boldsymbol{T}$$

The multiplication yields

$$P_j^{(n+1)} = \left(\frac{r - j + 1}{r} \right) P_j^{(n)} + \frac{1}{r} \sum_{i=j+1}^{r} P_i^{(n)} \tag{29}$$

Let us define

$$R_j^{(n)} \equiv \sum_{i=j+1}^{r} P_i^{(n)}$$

This is the probability that the nth car has a velocity greater than V_j. We have, from Eq. (28)

$$1 - R_j^{(n)} = \sum_{i=1}^{j} P_i^{(n)} = \sum_{i=1}^{j} \frac{(1 + r - j)^n - (r - j)^n}{r^n}$$

Then

$$R_j^{(n)} = 1 - \frac{1}{r^n}[(1 + r - 1)^n - (r - 1)^n + (1 + r - 2)^n - (r - 2)^n$$

$$+ \dots + (1 + r - j)^n - (r - j)^n] = 1 - \frac{1}{r^n}[r^n - (r - j)^n]$$

so
$$R_j^{(n)} = \left(\frac{r - j}{r} \right)^n \tag{30}$$

We substitute Eq. (28) and Eq. (30) into Eq. (29) and obtain

$$P_j^{(n+1)} = \frac{1}{r} \left\{ (1 + r - j) \frac{(1 + r - j)^n - (r - j)^n}{r^n} + \left(\frac{r - j}{r} \right)^n \right\}$$

$$= \frac{1}{r^{n+1}} \left\{ (1 + r - j)^{n+1} - (1 + r - j)(r - j)^n + (r - j)^n \right\}$$

$$= \frac{1}{r^{n+1}} \left\{ (1 + r - j)^{n+1} - (r - j)^{n+1} \right\}$$

This is the value of $P_j^{(n+1)}$ as given by Eq. (28).

References p. 237/238

On inspection of Eq. (28), it is seen that, for the higher velocity states, the occupancy probabilities drop off very rapidly with n. For example, the probability that the nth car in a platoon is in the highest state, $j = r$, is

$$P_r^{(n)} = \left(\frac{1}{r^n}\right)$$

Let us make a momentary digression and consider the continuous velocity spectrum corresponding to the discrete one we have used so far. Let V_1 and $V_r > V_1$ be the limits of a rectangular, continuous velocity density function. By letting r approach infinity in Eq. (30), we make a transition to the continuous case, obtaining

$$R^{(n)}(V) = \left(\frac{V_r - V}{V_r - V_1}\right)^n$$

which is the probability that the nth car in a platoon goes through the bottleneck at a velocity greater than V. While this formula is of interest, being susceptible to experimental comparison, we do not follow it up further because our concern is with the bottleneck's effect on flow. As we have noted before, the velocity distribution is a very indirect flow indicator.

A more suitable flow measure, as used in the simulation, is the average time headway of the nth car in the platoon. We resume the discrete velocity distribution approach. Then, since the segments $(V_{j+1} - V_j)$ are all the same, and since we have assumed a small overall velocity range, it is reasonable to assume that the relation between time headway and velocity near the beginning of the bottleneck is

$$\varDelta_j = (A - Bj) \tag{31}$$

for a vehicle of velocity V_j. Here, A and B are assumed constant. This is a crude model, but it has the right qualitative behavior.

The average time headway for the nth car is

$$\bar{\varDelta}^{(n)} = \sum_{j=1}^{r} \varDelta_j P_j^{(n)} = A - B \sum_{j=1}^{r} j P_j^{(n)}$$

$$= A - B \sum_{j=0}^{r-1} R_j^{(n)} = A - B \sum_{j=0}^{r-1} \left(\frac{r-j}{r}\right)^n \tag{32}$$

Assuming that r is large, we approximate

$$\bar{\varDelta}^{(n)} = A - B \int_0^{r-1} \left(\frac{r-x}{r}\right)^n dx \qquad \bar{\varDelta}^{(n)} = A - Br\left(\frac{1}{n+1}\right) \tag{33}$$

We compare this to the result from the computer model for trucks, all of which are affected by the bottleneck, as shown in Fig. 4. Setting the headways equal for $n = 1$ and $n = \infty$, we find

$$A = 2.52 \text{ seconds} \qquad Br = 0.84 \text{ seconds}$$

so

$$\bar{A}^{(n)} = 2.52 - 0.84 \left(\frac{1}{n + 1} \right) \text{ seconds} \qquad (34)$$

Fig. 13 compares this relation to that obtained with the computer simulation. Clearly, the agreement is not very good, since the simulation model gives, approximately, an exponential, rather than an inverse linear, dependence on n.

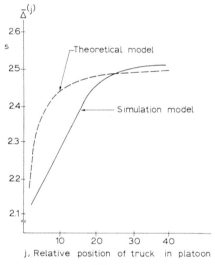

Fig. 13. Bottleneck average time headways, $\bar{A}^{(j)}$, as function of vehicle position within platoon : comparison of theoretical and computer model. The platoon consists entirely of trucks whose maximum acceleration is cut in half within the bottleneck which is 1024 ft. long. $---$ Theoretical model, Eq. (34). $\underline{\qquad}$ Simulation model, from Fig. 4.

While we have not gone beyond the formulation presented here, there is hope that the approach may become more realistic if we require that the arrival velocity distribution of the nth vehicle depends on that of the $(n - 1)$th. This is the actual situation, but such realism may make it very difficult to obtain a significant solution.

CONCLUDING REMARKS

The simulation model of this paper was applied to some problems beyond the scope of the present paper[6]. It was shown that the model drivers, in the aggregate, exhibit shock waves, a flow–density relation, and platooning in a fairly realistic manner. Nevertheless, some shortcomings became apparent.

The drivers' reactions to sharp maneuvers by a platoon leader were unsatisfactorily sluggish. This may be rectified by providing a variable velocity control C_1 and a variable reaction delay τ, as suggested by Fig. 1. When the model driver reacts to sharp maneuvers, C_1 should be large and τ should be small. On the other hand, in smooth traffic with weak or occasional maneuvers, C_1 should be small and τ should be large. A reasonable formulation might be the replacement of (C_1, τ) in Eq. (1), namely,

$$\ddot{x}_n (t + \tau) = C_1 [\dot{x}_{n-1}(t) - \dot{x}_n(t)] + \text{other terms}$$

by non-constant terms

$$C_1 = \frac{C_5}{x_{n-1}(t) - x_n(t)} \qquad \tau = \frac{1}{C_6 C_1}$$

where C_5 and C_6 are new constants.

It appears that our evaluation of the model parameters may well stand elaboration. They are too interdependent for really adequate data fitting by the simple method used here. In particular, the experimental reaction delay time is not entirely represented by its model counterpart. The model's effective delays are larger, being functions also of the sensitivity S and decision parameter K.

Nevertheless, the existing model has proved useful in clarifying bottleneck behavior. We expect that the deliberate platooning of vehicles, championed here, will be found a useful tool for maximizing flow in at least some congested roadways.

ACKNOWLEDGEMENTS

This study was conducted with the help of the Port of New York Authority. Mr. L. EDIE and his group at the Authority contributed practical advice and illuminating experimental data. IBM 704 Computer time was made available by the M.I.T. Computation Center. At M.I.T., Dr. H. GALLIHER and the author's thesis advisor, Professor P. M. MORSE, helped significantly in the formulation of the project.

REFERENCES

1 R. E. CHANDLER, R. HERMAN AND E. W. MONTROLL, Traffic Dynamics: Studies in Car Following, *Operat. Research*, 6 (1958) 165–184.
2 L. C. EDIE AND R. S. FOOTE, *Traffic Flow in Tunnels*, The Port of New York Authority, New York, 1957, and *Proc. Highway Research Board*, 37 (1958) 334–344.
3 W. FELLER, *Introduction to Probability Theory and its Applications*, Vol. 1, Wiley, New York, 2nd Ed., 1957, p. 380–383.

4 T. W. Forbes, M. J. Zagorski and W. A. Deterline, Measurement of Driver Reactions to Tunnel Conditions and Effects on Traffic Flow, *AIR–227–57–FR–157, Am. Inst. Research* Pittsburgh, Pa., July 10, 1957.
5 D. C. Gazis, R. Herman and R. B. Potts, Car-Following Theory of Steady-State Traffic Flow, *Operat. Research*, 7 (1959) 499–505.
6 W. Helly, Dynamics of Single-Lane Vehicular Traffic Flow, *Ph. D. Thesis*, Mass. Inst. Technol., 1959. Reprinted as *Research Rept. no. 2, Center for Operat. Research*, Mass. Inst. Technol., 1959.
7 O. K. Norman and W. P. Walker, *Highway Capacity Manual*, U. S. Dept. of Commerce, Bureau of Public Roads, 1950.
8 L. A. Pipes, An Operational Analysis of Traffic Dynamics, *J. Appl. Phys.*, 24 (1953) 274–281.
9 *Traffic Engineering Handbook*, Institute of Traffic Engineers, (New Haven, Conn.), 2nd Ed., 1950.
10 R. Herman, E. W. Montroll, R. B. Potts and R. W. Rothery, Traffic Dynamics: Analysis of Stability in Car Following, *Operat. Research*, 7 (1959) 86–106.